STUDENT'S SOLUTIONS

TO ACCOMPANY
JAMES T. McCLAVE AND FRANK H. DIETRICH, II'S

A FIRST COURSE IN STATISTICS

FOURTH
EDITION

NANCY S. BOUDREAU
Bowling Green State University

DELLEN PUBLISHING COMPANY
an imprint of
MACMILLAN PUBLISHING COMPANY
NEW YORK

MAXWELL MACMILLAN CANADA
TORONTO

Printed in the United States of America

Macmillan Publishing Company
866 Third Avenue, New York, New York 10022

Macmillan Publishing Company is
part of the Maxwell Communication
Group of Companies.

Maxwell Macmillan Canada, Inc.
1200 Eglinton Avenue East
Suite 200
Don Mills, Ontario M3C 3N1

Permissions: Dellen Publishing Company
400 Pacific Avenue
San Francisco, California 94133

Orders: Dellen Publishing Company
c/o Macmillan Publishing Company
Front and Brown Streets
Riverside, New Jersey 08075

ISBN: 0-02-312712-0

Printing: 1 2 3 4 5 6 Year: 2 3 4 5 6

Preface

This solutions manual is designed to accompany the text *A First Course in Statistics*, Fourth Edition, by James T. McClave and Frank H. Dietrich, II (Dellen Publishing Company, 1992). It provides answers to most odd-numbered exercises for each chapter in the text. Other methods of solution may also be appropriate; however, the author has presented one that she believes to be most instructive to the beginning statistics student. The student should first attempt to solve the assigned exercises without help from this manual. Then, if unsuccessful, the solution in the manual will clarify points necessary to the solution. The student who successfully solves an exercise should still refer to the manual's solution. Many points are clarified and expanded upon to provide maximum insight into and benefit from each exercise.

Instructors will also benefit from the use of this manual. It will save time in preparing presentations of the solutions and possibly provide another point of view regarding their meaning.

Some of the exercises are subjective in nature and thus omitted from the Answer Key at the end of *A First Course In Statistics*, Fourth Edition. The subjective decisions regarding these exercises have been made and are explained by the author. Solutions based on these decisions are presented; the solution to this type of exercise is often most instructive. When an alternative interpretation of an exercise may occur, the author has often addressed it and given justification for the approach taken.

I would like to thank Brenda Dobson for her assistance and for typing this work.

Nancy S. Boudreau
Bowling Green State University
Bowling Green, Ohio

Contents

CHAPTER

1

WHAT IS STATISTICS?

1.1 Descriptive statistics utilizes numerical and graphical methods to look for patterns, to summarize, and to present the information in a set of data. Inferential statistics utilizes sample data to make estimates, decisions, predictions, or other generalizations about a larger set of data.

1.3 A population is a set of existing units such as people, objects, transactions, or events. A variable is a characteristic or property of an individual population unit such as height of a person, time of a reflex, amount of a transaction, etc.

1.5 An inference without a measure of reliability is nothing more than a guess. A measure of reliability separates statistical inference from fortune telling or guessing. Reliability gives a measure of how confident one is that the inference is correct.

1.7 In order for the sample to be a random sample, every member of the class must have an equal chance of being selected for the sample.

1.9 a. The population of interest is all citizens of the United States, the same as in problem 1.8.

b. The variable of interest is the rating of the Presidential job performance from 0 to 100 by each citizen.

c. The sample is the 2000 individuals selected for the poll, the same as in problem 1.8.

d. The inference of interest might be the average job performance rating of the President by all citizens.

1.11 a. The population of interest is all possible 2-ounce portions of a basic solution.

b. The variable of interest is the amount of hydrochloric acid necessary to neutralize the 2 ounces of basic solution.

c. The sample consists of the 5 2-ounce portions of the solution prepared.

1.13 a. The population of interest is all medical doctors.

b. The variable of interest is whether or not the doctor has been involved in one or more malpractice suits.

c. The sample consists of the 500 medical doctors randomly selected.

d. The inference of interest is to estimate the proportion of medical doctors who have been involved in one or more malpractice suits.

1.15 a. The population of interest is the collection of all department store executives.

b. Two variables are measured. The first is the job satisfaction of each executive and the second is the "Machiavellian" rating of each executive.

c. The sample is the collection of the 218 department store executives selected for the study.

d. The authors concluded that those executives with higher job satisfaction scores are likely to have a lower "Mach" rating.

1.17 a. One population of interest is the set of all elderly people who have been treated by the new method of treating blindness. The other population of interest is the set of all elderly people who have not been treated.

b. One sample consists of the 224 patients treated with the new method. The other sample consists of the selected group of untreated elderly people.

c. The variable of interest is the condition of the patients (blind or not).

d. The researchers want to know if the new laser beam treatment was effective. They will compare the proportions of elderly people who go blind in each group. The inference will be based on the sample proportions.

CHAPTER

2

METHODS FOR DESCRIBING SETS OF DATA

2.1 a. Nominal data are measurements that simply classify the units of the sample or population into categories. These categories cannot be ranked. Ordinal data are measurements that enable the units of the sample or population to be ordered or ranked with respect to the variable of interest.

b. Interval data are measurements that enable the determination of the differential (how much more or less) of the characteristic being measured between one unit of the sample or population and another. Interval data will always be numerical, and the numbers assigned to the two units can be subtracted to determine the difference between the units. However, the zero point is not meaningful for these data. Thus, these data cannot be multiplied or divided. Ratio data are measurements that enable the determination of the multiple (how many times as much) of the characteristic being measured between one unit of the sample or population and another. All the characteristics of interval data are included in ratio data. In addition, the zero point for ratio data is meaningful

c. Qualitative data have no meaningful numbers associated with them. Qualitative data include nominal and ordinal data. Quantitative data have meaningful numbers associated with them. Quantitative data include interval and ratio data.

2.3 The data consisting of the classifications A, B, C, and D are qualitative. These data are nominal and thus are qualitative. After the data are input as 1, 2, 3, and 4, they are still nominal and thus qualitative. The only difference between the two data sets are the names of the categories. The numbers associated with the four groups are meaningless.

2.5 a. Nominal; possible brands are "Guess," "Levis," "Lee," etc., each of which represents a nonranked category.

b. Ratio; the number of hours of sports programming carried in a typical week is measured on a numerical scale where the zero point has meaning. Ten hours of sports programming is five times as much as two hours of sports programming.

c. Ratio; the percentage of their time spent studying is measured on a numerical scale where the zero point has meaning. Forty percent of time spent studying is twice as much as twenty percent.

d. Ratio; the number of long distance telephone calls is measured on a numerical scale where the zero point has meaning. Twenty phone calls is four times as many as five phone calls.

e. Interval; SAT scores are measured on a numerical scale where the zero point has no meaning. A score of 500 is not twice as good as a score of 250.

2.7 a. Nominal; brand of stereo speaker would be measured using nonranked categories.

b. Ratio; loss (in dollars) is measured on a numerical scale where the zero point has meaning. A loss of $600 is three times as much as a loss of $200.

c. Nominal; color is measured using nonranked categories.

d. Ordinal; ranking of football teams orders the teams, but does not indicate how much better one team is than another.

2.9 The stem will consist of the digits to the left of the decimal point, while the leaves will consist of all the digits to the right of the decimal point. The stem and leaf display is:

Stem	Leaf
0	8
1	1 6 7 9
2	6 4 8 6 0 9 9 5
3	3 5 9 4
4	1 5
5	0

2.11 The stem consists of all the digits at or to the left of the 10's digit, while the leaves consist of all the digits to the right of the 10's digit. However, only one digit is generally displayed in the leaf, so the decimal portion of the numbers is dropped. The stem and leaf display is:

Stem	Leaf
0	8 9
1	7 6 2 9 0
2	0 9 8 4 0 9 3 2 1
3	2 3
4	2
5	2

Most of the observations are between 10 and 30. Only two observations are below 10, while only four are above 30.

2.13 a. The stems will consist of the digits in the 10's column, while the leaves will consist of all the digits to the right of the 10's column. However, only one digit is generally displayed in the leaf, so the decimal portion of the number is dropped.

1982

Stem	Leaf
1	5 9 1 8 6 7 6
2	6 9 9 5 5 5 3 8 7 5 3 8 5 1 3 3 2 9 7 4 7 5 0 6 3 7
3	6 5 6 9 9 5 4 8 8 5 0 6 2 7 2 6 3
4	3

1984

Stem	Leaf
1	4 8 0 7 3 3 4 6 5
2	5 4 4 0 8 6 4 5 3 2 2 5 7 3 4 2 9 0 6 6 2 9 1 5 4 6 4
3	7 5 6 7 6 1 7 3 7 0 1 5 5
4	4 3

b. In 1982, the highest dropout rate was 43.1% (D.C.), while the lowest dropout rate was 11.8% (Minnesota).

In 1984, the highest dropout rate was 44.8% (D.C.), while the lowest dropout rate was 10.7% (Minnesota).

The states with the highest and lowest dropout rates remained the same in 1982 and 1984.

c. For 1982, the middle or 26th dropout rate is 27.6%, while the middle dropout rate in 1984 is 25.3%. The "middle" dropout rate has decreased 2.3%, from 27.6% to 25.3%.

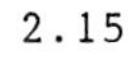

2.15

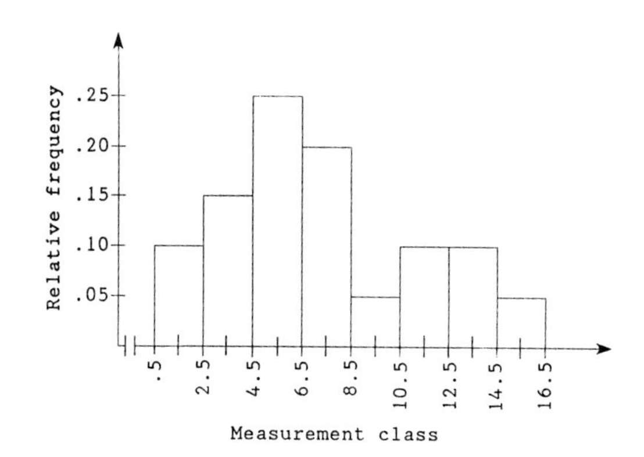

2.17 The smallest value in the data set is 12, and the largest is 99; thus, the range to be spanned is 87. We choose to use 8 classes with a class width of

$$\frac{87}{8} \approx 11$$

To generate the classes, we will begin at 11.5 and successively add 11, ending at 99.5.

MEASUREMENT CLASS	FREQUENCY	RELATIVE FREQUENCY
11.5-22.5	8	8/30 = .267
22.5-33.5	5	5/30 = .167
33.5-44.5	4	.133
44.5-55.5	6	.200
55.5-66.5	2	.067
66.6-77.5	1	.033
77.5-88.5	3	.100
88.5-99.5	1	.033

The corresponding relative frequency histogram is shown below.

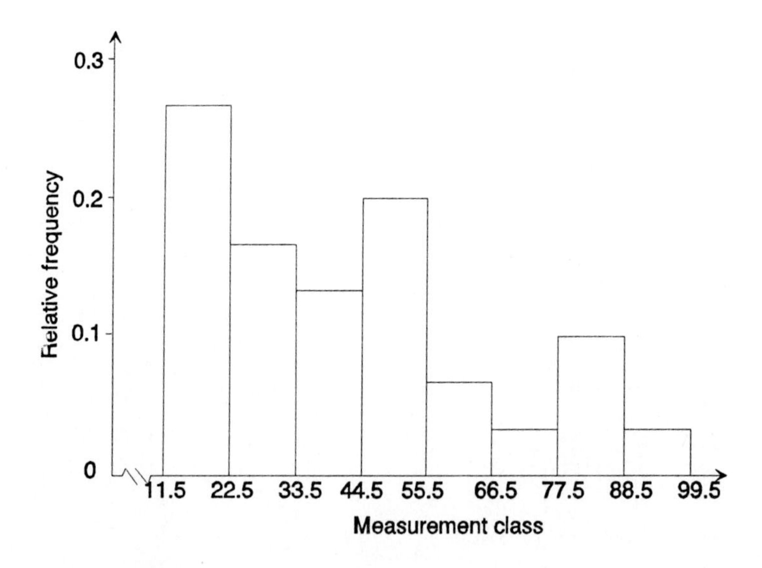

2.19 The range of the n = 20 data points is 52.8 - 8.4 = 44.4. We choose to use 5 classes of width

$$\frac{44.4}{5} \approx 9$$

beginning at 8.35.

MEASUREMENT CLASS	FREQUENCY	RELATIVE FREQUENCY
8.35-17.35	5	.25
17.35-26.35	8	.40
26.35-35.35	5	.25
35.35-44.35	1	.05
44.35-53.35	1	.05

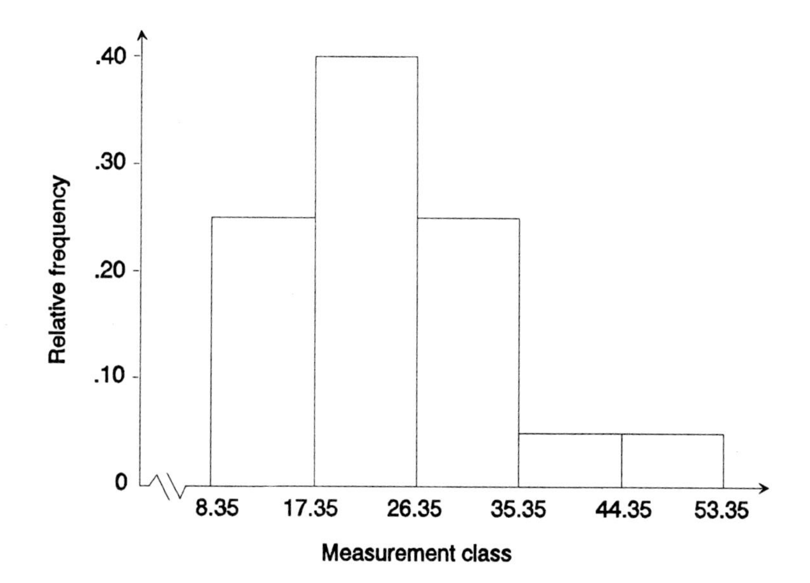

The stem and leaf display uses the digits in the tens place as stems and is shown below. Generally, only one digit is used for the leaf, so the decimal portion of the numbers is dropped.

Stem	Leaf
0	8 9
1	6 0 2 7 9
2	9 8 3 4 2 0 1 0 9
3	2 3
4	2
5	2

Note the similarities between the relative frequency histogram and the stem and leaf display.

2.21 a. The class width is $\frac{21.45 - .45}{6} = 3.5$. The frequency table is:

MEASUREMENT CLASS	FREQUENCY	RELATIVE FREQUENCY
.45- 3.95	7	.35
3.95- 7.45	8	.40
7.45-10.95	2	.10
10.95-14.45	1	.05
14.45-17.95	0	.00
17.95-21.45	2	.10

The relative frequency histogram is:

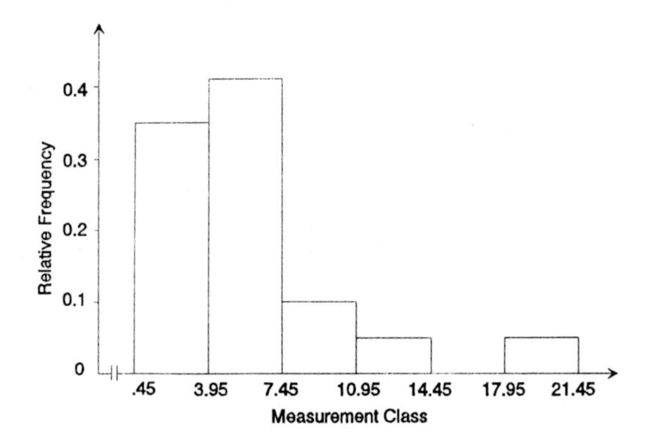

b. There are 3 observations greater than 13. Therefore, the proportion eligible for benefits is 3/20 = .15.

2.23 a. We use 6 classes of length 4, commencing with 5.5.

MEASUREMENT CLASS	FREQUENCY	RELATIVE FREQUENCY
5.5- 9.5	2	.10
9.5-13.5	7	.35
13.5-17.5	7	.35
17.5-21.5	2	.10
21.5-25.5	1	.05
25.5-29.5	1	.05

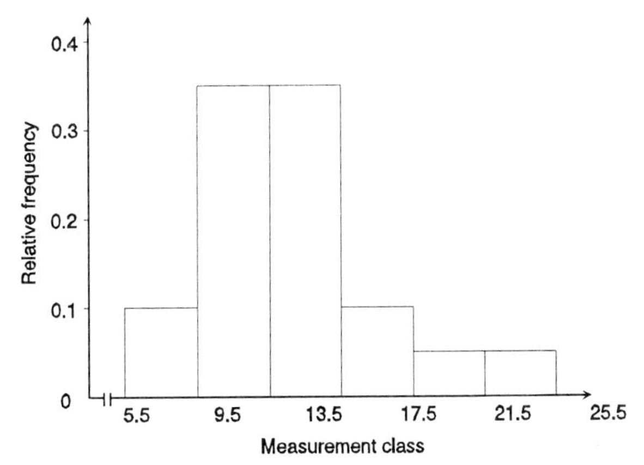

b. From the original data set, the proportion of years in which pots have to be lit 15 days or less is 13/20 = .65.

2.25 a. There are 30 observations in each set of data, so we (arbitrarily) choose to use 6 classes. To determine the class width, we take the difference between the largest and smallest measurement in the data set divided by the number of classes. Since we want to use the same class width for both data sets, we will use the largest difference to determine class width (from 1980 data).

$$\text{Class width} = \frac{1595 - 128}{6} = \frac{1467}{6} = 244.5 \approx 245$$

The first measurement class will begin .5 below the smallest measurement or 10.5. The relative frequency distributions for the two data sets are:

1960

MEASUREMENT CLASS	FREQUENCY	RELATIVE FREQUENCY
10.5-255.5	16	16/30 = .533
255.5-500.5	8	8/30 = .267
500.5-745.5	5	5/30 = .167
745.5-990.5	1	1/30 = .033
	n = 30	1.000

1980		
MEASUREMENT CLASS	FREQUENCY	RELATIVE FREQUENCY
10.5- 255.5	10	10/30 = .333
255.5- 500.5	11	11/30 = .367
500.5- 745.5	4	4/30 = .133
745.5- 990.5	4	4/30 = .333
990.5-1235.5	0	0/30 = 0
1235.5-1480.5	0	0/30 = 0
1480.5-1725.5	1	1/30 = .033
		.999

The histogram for 1960 is:

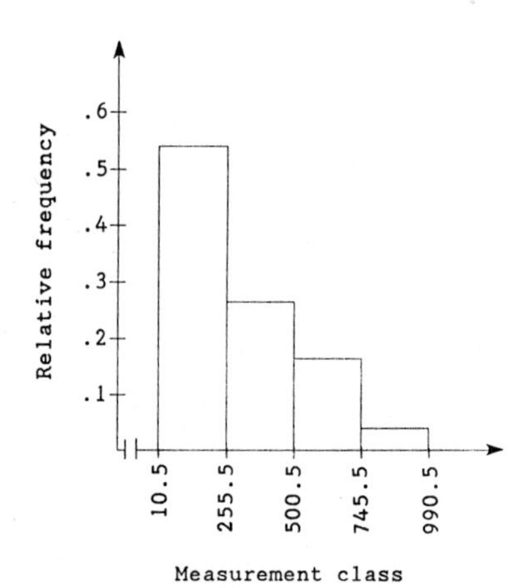

The histogram for the 1980 data is:

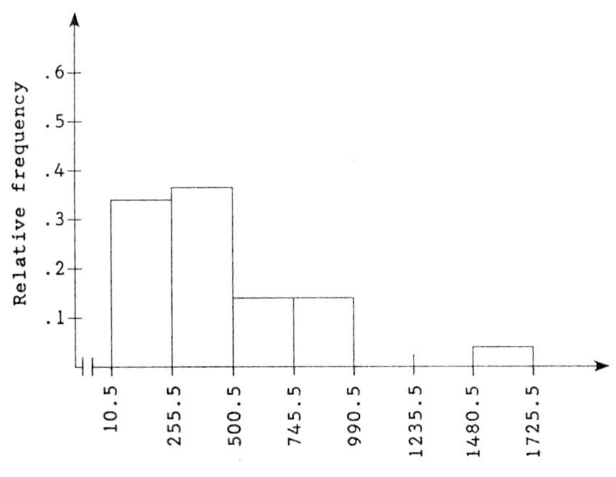

b. The population appears to be shifting to the right. In 1980, the relative frequency in the lowest class has decreased since 1960, while the relative frequency has increased in the other classes.

2.27 Assume the data are a sample. The mode is the observation that occurs most frequently. For this sample, the mode is 15, which occurs 3 times.

The sample mean is:

$$\bar{x} = \frac{\sum x_i}{n} = \frac{18 + 10 + 15 + 13 + 17 + 15 + 12 + 15 + 18 + 16 + 11}{11}$$

$$= \frac{160}{11} = 14.545$$

The median is the middle number when the data are arranged in order. The data arranged in order are: 10, 11, 12, 13, 15, 15, 15, 16, 17, 18, 18. The middle number is the 6th number, which is 15.

2.29 The median is the middle number once the data have been arranged in order. If n is even, there is not a single middle number. Thus, to compute the median, we take the average of the middle two numbers. If n is odd, there is a single middle number. The median is this middle number.

A data set with 5 measurements arranged in order is 1, 3, 5, 6, 8. The median is the middle number, which is 5.

A data set with 6 measurements arranged in order is 1, 3, 5, 5, 6, 8. The median is the average of the middle two numbers which is

$$\frac{5 + 5}{2} = \frac{10}{2} = 5$$

2.31 a. $\bar{x} = \frac{\sum x}{n} = \frac{85}{10} = 8.5$

b. $\bar{x} = \frac{400}{16} = 25$

c. $\bar{x} = \frac{35}{45} = .78$

d. $\bar{x} = \frac{242}{18} = 13.44$

2.33 a. For a distribution that is skewed to the left, the mean is less than the median.

b. For a distribution that is skewed to the right, the mean is greater than the median.

c. For a symmetric distribution, the mean and median are equal.

2.35 b. The relative frequency histogram will be based upon the following frequency table, which utilizes six measurement classes.

MEASUREMENT CLASS	FREQUENCY	RELATIVE FREQUENCY
63.5-69.5	3	3/20 = .15
69.5-75.5	3	.15
75.5-81.5	3	.15
81.5-87.5	6	.30
87.5-93.5	4	.20
93.5-99.5	1	.05
	n = 20	

The relative frequency histogram is shown below.

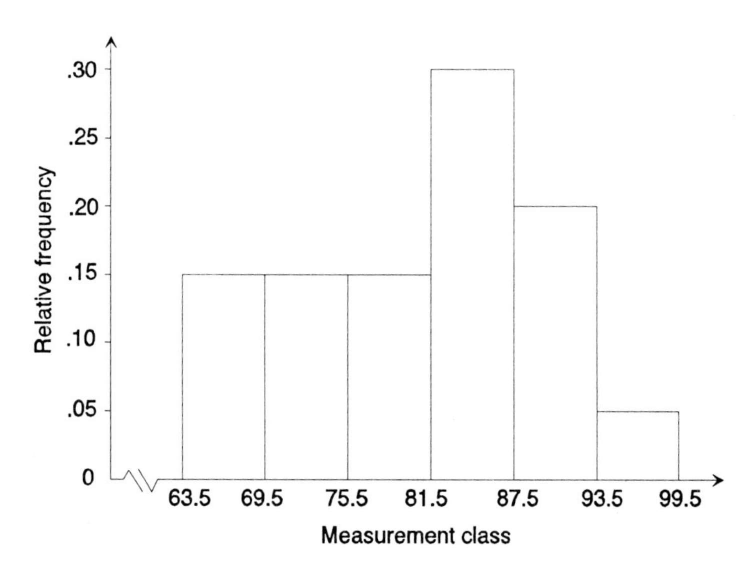

c. We first rank the data in ascending order:

64	72	80	84	90
66	75	83	86	91
68	79	83	87	93
71	80	83	90	98

The following summary statistics can now be computed:

mean: $\bar{x} = \frac{\sum x}{n} = \frac{1623}{20} = 81.15$

median: the average of the middle (10th and 11th) measurements

$$\text{median} = \frac{83 + 83}{2} = 83$$

mode: most frequently occurring value

mode = 83

These measures seem to reflect the central tendency of the data set.

2.37 a. The mean is $\bar{x} = \frac{\sum x}{n} = \frac{290 + 90 + \ldots + 110}{19} = \frac{3420}{19} = 180$

To find the median, we first arrange the data in order from the smallest to the largest. Because n is odd, the median will be the middle or 10th data point. For this set of data, the median is 162.

b. Because the mean is larger than the median, the data set is skewed to the right. There are a few large numbers that pull the mean up.

c. If n is odd, the median will always be an actual value in the data set. If n is even, the median will be equal to an actual value only if the middle two numbers are equal to each other.

2.39 a. Since two rats did not yield an escape time before the experiment was terminated, we cannot compute the mean escape time for the ten rats involved in the experiment.

b. It is appropriate to compute the median, because the two N's represent extremely large times and have no effect on the calculation of the median. The ordered data are:

38 56 95 100 104 116 122 135 N N

Then, median $= \frac{104 + 116}{2} = 110$

2.41 The total amount saved was \$2.24 billion. Thus, $\sum x_i = 2.24$. The total number, n, of coupons redeemed was 6.49 billion. The mean is

$$\bar{x} = \frac{\sum x}{n} = \frac{2.24}{6.49} = .345 \approx \$.35$$

2.43 a. The mean or average dropout rate in 1982 was 28.12%, while the mean or average dropout rate in 1984 was 26.50%. Thus, the mean dropout rate has dropped from 1982 to 1984.

The median or middle dropout rate for 1982 is 27.60%, while the median or middle dropout rate for 1984 is 25.30%. Again, there has been a drop in the median dropout rate from 1982 to 1984.

b. In both years, the mean is larger than the median. This implies that in both years, the data sets are skewed to the right. This is supported by both the stem and leaf displays and the histograms.

c. The distribution appears to have shifted from right to left. Because the sample sizes were large in both years (n = 51), it appears that the shift is indicative of some trend and not some random variation. Also, there was a decrease in the dropout rate for 39 of the 51 states.

2.45 The sample variance is the sum of the squared deviations from the sample mean divided by the sample size minus 1. The population variance is the sum of the squared deviations from the population mean divided by the population size.

2.47 a. Range = 4 - 0 = 4

$$s^2 = \frac{1}{n - 1}\{\Sigma x^2 - \frac{(\Sigma x)^2}{n}\} = \frac{1}{4}\{22 - \frac{(8)^2}{5}\} = 2.30$$

$$s = \sqrt{s^2} = \sqrt{2.30} = 1.52$$

b. Range = 6 - 0 = 6

$$s^2 = \frac{1}{n - 1}\{\Sigma x^2 - \frac{(\Sigma x)^2}{n}\} = \frac{1}{6}\{63 - \frac{(17)^2}{7}\} = 3.62$$

$$s = \sqrt{s^2} = \sqrt{3.62} = 1.90$$

c. Range = 8 - (-2) = 10

$$s^2 = \frac{1}{n - 1}\{\Sigma x^2 - \frac{(\Sigma x)^2}{n}\} = \frac{1}{9}\{154 - \frac{(30)^2}{10}\} = 7.11$$

$$s = \sqrt{s^2} = \sqrt{7.11} = 2.67$$

d. Range = 2 - (-3) = 5

$$s^2 = \frac{1}{n - 1}\{\Sigma x^2 - \frac{(\Sigma x)^2}{n}\} = \frac{1}{17}\{29 - \frac{(-5)^2}{18}\} = 1.62$$

$$s = \sqrt{s^2} = \sqrt{1.62} = 1.27$$

2.49 a. Range = 42 - 37 = 5

$$s^2 = \frac{1}{n - 1}\{\Sigma x^2 - \frac{(\Sigma x)^2}{n}\} = \frac{1}{4}\{7935 - \frac{(199)^2}{5}\} = 3.70$$

$$s = \sqrt{s^2} = \sqrt{3.70} = 1.92$$

b. Range = 100 - 1 = 99

$$s^2 = \frac{1}{n - 1}\{\Sigma x^2 - \frac{(\Sigma x)^2}{n}\} = \frac{1}{8}\{25795 - \frac{(303)^2}{9}\} = 1949.25$$

$$s = \sqrt{s^2} = \sqrt{1949.25} = 44.15$$

c. Range = 100 - 2 = 98

$$s^2 = \frac{1}{n - 1}\{\Sigma x^2 - \frac{(\Sigma x)^2}{n}\} = \frac{1}{7}\{20033 - \frac{(295)^2}{8}\} = 1307.84$$

$$s = \sqrt{s^2} = \sqrt{1307.84} = 36.16$$

2.51 a. $$\sum_{i=1}^{n} x_i = 3 + 1 + 10 + 10 + 4 = 28$$

$$\sum_{i=1}^{n} x_i^2 = 3^2 + 1^2 + 10^2 + 10^2 + 4^2 = 226$$

$$\bar{x} = \frac{\sum_{i=1}^{n} x_i}{n} = \frac{28}{5} = 5.6$$

$$s^2 = \frac{\sum_{i=1}^{n} x_i^2 - \frac{\left(\sum_{i=1}^{n} x_i\right)^2}{n}}{n - 1} = \frac{226 - \frac{28^2}{5}}{5 - 1} = \frac{69.2}{4} = 17.3$$

$$s = \sqrt{17.3} = 4.1593$$

b. $$\sum_{i=1}^{n} x_i = 8 + 10 + 32 + 5 = 55$$

$$\sum_{i=1}^{n} x_i^2 = 8^2 + 10^2 + 32^2 + 5^2 = 1213$$

$$\bar{x} = \frac{\sum_{i=1}^{n} x_i}{n} = \frac{55}{4} = 13.75 \text{ feet}$$

$$s^2 = \frac{\sum_{i=1}^{n} x_i^2 - \frac{\left(\sum_{i=1}^{n} x_i\right)^2}{n}}{n - 1} = \frac{1213 - \frac{55^2}{4}}{4 - 1} = \frac{456.75}{3} = 152.25 \text{ square feet}$$

$$s = \sqrt{152.25} = 12.339 \text{ feet}$$

c. $$\sum_{i=1}^{n} x_i = -1 + (-4) + (-3) + 1 + (-4) + (-4) = -15$$

$$\sum_{i=1}^{n} x_i^2 = (-1)^2 + (-4)^2 + (-3)^2 + 1^2 + (-4)^2 + (-4)^2 = 59$$

$$\bar{x} = \frac{\sum_{i=1}^{n} x_i}{n} = \frac{-15}{6} = -2.5$$

$$s^2 = \frac{\sum_{i=1}^{n} x_i^2 - \frac{\left(\sum_{i=1}^{n} x_i\right)^2}{n}}{n - 1} = \frac{59 - \frac{(-15)^2}{6}}{6 - 1} = \frac{21.5}{5} = 4.3$$

$$s = \sqrt{4.3} = 2.0736$$

d. $$\sum_{i=1}^{n} x_i = \frac{1}{5} + \frac{1}{5} + \frac{1}{5} + \frac{2}{5} + \frac{1}{5} + \frac{4}{5} = \frac{10}{5} = 2$$

$$\sum_{i=1}^{n} x_i = \left(\frac{1}{5}\right)^2 + \left(\frac{1}{5}\right)^2 + \left(\frac{1}{5}\right)^2 + \left(\frac{2}{5}\right)^2 + \left(\frac{1}{5}\right)^2 + \left(\frac{4}{5}\right)^2 = \frac{24}{25} = .96$$

$$\bar{x} = \frac{\sum_{i=1}^{n} x_i}{n} = \frac{2}{6} = \frac{1}{3} = .33 \text{ ounce}$$

$$s^2 = \frac{\sum_{i=1}^{n} x_i^2 - \frac{\left(\sum_{i=1}^{n} x_i\right)^2}{n}}{n - 1} = \frac{\frac{24}{25} - \frac{2^2}{6}}{6 - 1} = \frac{.2933}{5} = .0587 \text{ square ounce}$$

$$s = \sqrt{.0587} = .2422 \text{ ounce}$$

2.53 This is one possibility for the two data sets.

Data Set 1: 0, 1, 2, 3, 4, 5, 6, 7, 8, 9

Data Set 2: 0, 0, 1, 1, 2, 2, 3, 3, 9, 9

The two sets of data above have the same range = largest measurement - smallest measurement = 9 - 0 = 9.

The means for the two data sets are:

$$\bar{x}_1 = \frac{\sum_{i=1}^{n} x_i}{n} = \frac{0 + 1 + 2 + 3 + 4 + 5 + 6 + 7 + 8 + 9}{10} = \frac{45}{10} = 4.5$$

$$\bar{x}_2 = \frac{\sum_{i=1}^{n} x_i}{n} = \frac{0 + 0 + 1 + 1 + 2 + 2 + 3 + 3 + 9 + 9}{10} = \frac{30}{10} = 3$$

The dot diagrams for the two data sets are shown below.

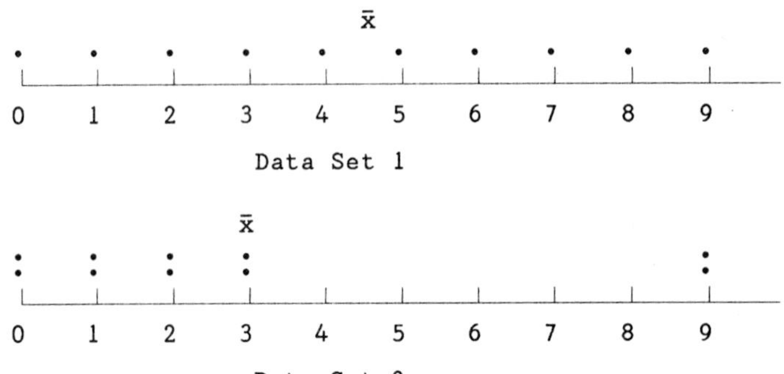

2.55 The first professor would probably be preferred because of the lower standard deviation of grade-point averages. Students in the first professor's class are less likely to have extremely low grade-point averages than are students in the second professor's class.

2.57 a. Germany - The range is 7.9 - 6.5 = 1.4
Italy - The range is 26.5 - 21.8 = 4.7
France - The range is 23.2 - 19.1 = 4.1
United Kingdom - The range is 20.1 - 11.3 = 8.8
Belgium - The range is 12.3 - 10.8 = 1.5

b. Germany

$$s^2 = \frac{\sum_{i=1}^{n} x_i^2 - \frac{\left(\sum_{i=1}^{n} x_i\right)^2}{n}}{n-1} = \frac{157.95 - \frac{21.7^2}{3}}{3-1} = .4933$$

$$s = \sqrt{.4933} = .7024$$

Italy

$$s^2 = \frac{\sum_{i=1}^{n} x_i^2 - \frac{\left(\sum_{i=1}^{n} x_i\right)^2}{n}}{n-1} = \frac{1748.7 - \frac{72.2^2}{3}}{3-1} = 5.5433$$

$$s = \sqrt{5.5433} = 2.3544$$

France

$$s^2 = \frac{\sum_{i=1}^{n} x_i^2 - \frac{\left(\sum_{i=1}^{n} x_i\right)^2}{n}}{n - 1} = \frac{1365.8 - \frac{63.8^2}{3}}{3 - 1} = 4.2433$$

$$s = \sqrt{4.2433} = 2.0599$$

United Kingdom

$$s^2 = \frac{\sum_{i=1}^{n} x_i^2 - \frac{\left(\sum_{i=1}^{n} x_i\right)^2}{n}}{n - 1} = \frac{778.19 - \frac{47.1^2}{3}}{3 - 1} = 19.36$$

$$s = \sqrt{19.36} = 4.4$$

Belgium

$$s^2 = \frac{\sum_{i=1}^{n} x_i^2 - \frac{\left(\sum_{i=1}^{n} x_i\right)^2}{n}}{n - 1} = \frac{402.49 - \frac{34.7^2}{3}}{3 - 1} = .5633$$

$$s = \sqrt{.5633} = .7506$$

c. Using the ranges from part (a), the ranks of the countries are:

1. Germany (range = 1.4)
2. Belgium (range = 1.5)
3. France (range = 4.1)
4. Italy (range = 4.7)
5. United Kingdom (range = 8.8)

Using the standard deviations from part (b), the ranks of the countries are:

1. Germany (s = .7024)
2. Belgium (s = .7506)
3. France (s = 2.0599)
4. Italy (s = 2.3544)
5. United Kingdom (s = 4.4)

d. Yes, the two sets of rankings in part (c) are in agreement. But no, the range and standard deviation will not always yield the same rankings. A data set could have a large range due to one extreme value. Another data set could have a slightly smaller range but more values near the extreme, thus having a larger standard deviation.

2.59 a. The units of measurement of the variable of interest is dollars (the same as the mean and standard deviation). Based on this, the data are ratio. Data measured in dollars are measured on a numerical scale where the zero point has meaning.

b. Since no information is given about the shape of the data set, we can only use Chebyshev's theorem.

$900 is 2 standard deviations below the mean, and $2100 is 2 standard deviations above the mean. Using Chebyshev's theorem, at least 3/4 of the measurements (or $3/4 \times 200 = 150$ measurements) will fall between $900 and $2100.

$600 is 3 standard deviations below the mean and $2400 is 3 standard deviations above the mean. Using Chebyshev's theorem, at least 8/9 of the measurements (or $8/9 \times 200 \approx 178$ measurements) will fall between $600 and $2400.

$1200 is 1 standard deviation below the mean and $1800 is 1 standard deviation above the mean. Using Chebyshev's theorem, nothing can be said about the number of measurements that will fall between $1200 and $1800.

$1500 is equal to the mean and $2100 is 2 standard deviations above the mean. Using Chebyshev's theorem, at least 3/4 of the measurements (or $3/4 \times 200 = 150$ measurements) will fall between $900 and $2100. It is possible that all of the 150 measurements will be between $900 and $1500. Thus, nothing can be said about the number of measurements between $1500 and $2100.

2.61 According to the Empirical Rule:

a. Approximately 68% of the measurements will be contained in the interval $\bar{x} - s$ to $\bar{x} + s$.

b. Approximately 95% of the measurements will be contained in the interval $\bar{x} - 2s$ to $\bar{x} + 2s$.

c. Essentially all the measurements will be contained in the interval $\bar{x} - 3s$ to $\bar{x} + 3s$.

2.63 Using Chebyshev's theorem, at least 8/9 of the measurements will fall within 3 standard deviations of the mean. Thus, the range of the data would be around 6 standard deviations. Using the Empirical Rule, approximately 95% of the observations are within 2 standard deviations of the mean. Thus, the range of the data would be around 4 standard deviations. We would expect the standard deviation to be somewhere between Range/6 and Range/4.

For our data, the range = 760 - 135 = 625.

The Range/6 = 625/6 = 104.17 and Range/4 = 625/4 = 156.25.

Therefore, I would estimate that the standard deviation of the data set is between 104.17 and 156.25.

It would not be feasible to have a standard deviation of 25. If the standard deviation were 25, the data would span 625/25 = 25 standard deviations. This would be extremely unlikely.

2.65 a. To decide if Chebyshev's theorem or the Empirical Rule is more appropriate, we need to know if the sample has a mound-shaped distribution.

To construct a relative frequency histogram for the data set, we will use 7 measurement classes.

$$\text{The class interval width} = \frac{\text{Largest measurement - Smallest measurement}}{\text{Number of intervals}}$$

$$= \frac{290{,}000 - 65{,}000}{7} = 32142.857$$

Rounding upward, the class width is 32150.

The measurement classes, class frequencies, and class relative frequencies are shown in the following table.

CLASS	MEASUREMENT CLASS	CLASS FREQUENCY	CLASS RELATIVE FREQUENCY
1	64,999 - 97,149	26	.722
2	97,149 - 129,299	5	.139
3	129,299 - 161,449	1	.028
4	161,449 - 193,599	3	.083
5	193,599 - 225,749	0	.000
6	225,749 - 257,899	0	.000
7	257,899 - 290,049	1	.028
		n = 36	1.000

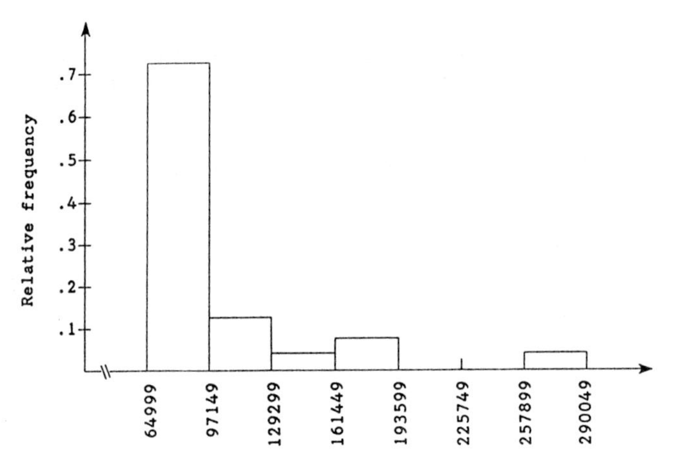

The relative frequency histogram above is not mound-shaped. Therefore, Chebyshev's theorem would be more appropriate.

b. The average cost of a new 1600-square-foot home with 1-1/2 or 2 baths in a desirable neighborhood for the 36 cities sampled is \$96,732.

The standard deviation is a measure of variability for the data set. For this sample, the standard deviation is \$45,322.

c. $\bar{x} \pm 2s \Rightarrow 96{,}732 \pm 2(45{,}322) \Rightarrow (6{,}088,\ 187{,}376)$

Using Chebyshev's theorem, we would expect at least $1 - \frac{1}{2^2} = .75$ of the sale prices to fall within this interval. For this data set, $\frac{34}{36} = .944$ of the sales prices actually fall in the interval. This is at least .75.

$\bar{x} \pm 3s \Rightarrow 96{,}732 \pm 3(45{,}322) \Rightarrow (-39{,}234,\ 232{,}698)$

Using Chebyshev's theorem, we would expect at least $1 - \frac{1}{3^2} = .889$ of the sale prices to fall within this interval. Thus, at most $1 - .889 = .111$ of the sale prices will fall outside this interval.

For this data set, $\frac{1}{36} = .028$ of the sale prices actually fall outside the interval. This is at most .111.

d. The sale price of a home in Darien, Connecticut, is \$290,000. The number of standard deviations \$290,000 is from the mean is

$$\frac{\$290{,}000 - 96732}{45322} = 4.26$$

2.67 Of the students who had a low level of math anxiety and were taught using a traditional expository method, approximately 68% should score within one standard deviation of the mean (131.16, 235.70); approximately 95% should score within two standard deviations of the mean (78.89, 287.97); and almost all should score within three standard deviations of the mean (26.62, 340.24).

2.69 Since we do not know if the distribution of the heights of the trees is mound-shaped, we need to apply Chebyshev's theorem. We know $\mu = 30$ and $\sigma = 3$. Therefore,

$$\mu \pm 3\sigma \Rightarrow 30 \pm 3(3) \Rightarrow 30 \pm 9 \Rightarrow (21, 39)$$

According to Chebyshev's theorem, at least $1 - \frac{1}{3^2} = \frac{8}{9}$ or .89 of the tree heights on this piece of land fall within this interval and at most $\frac{1}{9}$ or .11 of the tree heights will fall above the interval. However, the buyer will only purchase the land if at least $\frac{1000}{5000}$ or .20 of the tree heights are at least 40 feet tall. Therefore, the buyer should not buy the piece of land.

2.71 $\bar{x} = 125$, $s = 10$

a. $\bar{x} \pm 3s \Rightarrow 125 \pm 3(10) \Rightarrow 125 \pm 30 \Rightarrow (95, 155)$

Since nothing is known about the distribution of utility bills, we need to apply Chebyshev's Theorem. According to Chebyshev's theorem, the fraction of all three-bedroom homes with gas or electric energy that have bills within this interval is at least

$$1 - \frac{1}{3^2} = \frac{8}{9}$$

b. If it is reasonable to assume the distribution of utility bills is mound-shaped, we can apply the Empirical Rule.

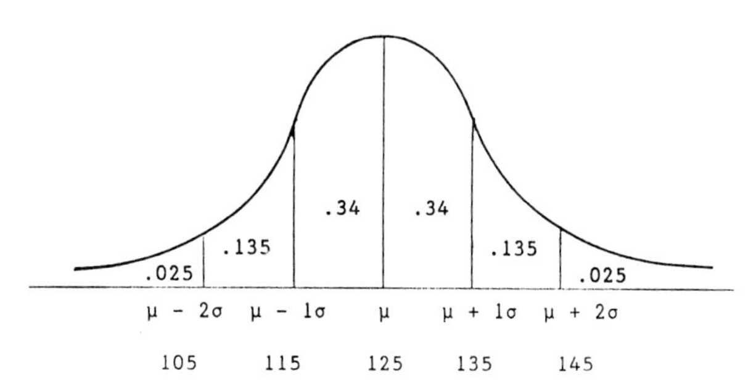

As illustrated in the figure, approximately .34 + .34 + .135 + .025 = .84 of three-bedroom homes would have monthly bills less than \$135.

c. Yes, these three values do suggest that solar energy units might result in lower utility bills. This is evident since if the solar energy units do not decrease the utility bills, only approximately .025 of the utility bills would be under \$105. But all three of the sampled bills from houses with solar energy units were under \$105.

2.73 At least 3/4 (4.5 => 5) of the athletes' performance times are within two standard deviations of the mean (21.30, 102.50), while at least 8/9 (5.33 => 6 = all) are within three standard deviations of the mean (1.00, 122.80).

2.75 Using the definition of a percentile:

	PERCENTILE	PERCENTAGE ABOVE	PERCENTAGE BELOW
a.	75th	25%	75%
b.	50th	50%	50%
c.	20th	80%	20%
d.	84th	16%	84%

2.77 We first compute z-scores for each x-value.

a. $z = \frac{x - \mu}{\sigma} = \frac{100 - 50}{25} = 2$

b. $z = \frac{x - \mu}{\sigma} = \frac{1 - 4}{1} = -3$

c. $z = \frac{x - \mu}{\sigma} = \frac{0 - 200}{100} = -2$

d. $z = \frac{x - \mu}{\sigma} = \frac{10 - 5}{3} = 1.67$

The above z-scores indicate that the x value in part (a) lies the greatest distance above the mean and the x value if part (b) lies the greatest distance below the mean.

2.79 a. From the problem, $\mu = 2.7$ and $\sigma = .5$

$$z = \frac{x - \mu}{\sigma} \Rightarrow z\sigma = x - \mu \Rightarrow x = \mu + z\sigma$$

For $z = 2.0$, $x = 2.7 + 2.0(.5) = 3.7$

For $z = -1.0$, $x = 2.7 - 1.0(.5) = 2.2$

For $z = .5$, $x = 2.7 + .5(.5) = 2.95$

For $z = -2.5$, $x = 2.7 - 2.5(.5) = 1.45$

b. For $z = -1.6$, $x = 2.7 - 1.6(.5) = 1.9$

c. If we assume the distribution of GPAs is approximately mound shaped, we can use the Empirical Rule.

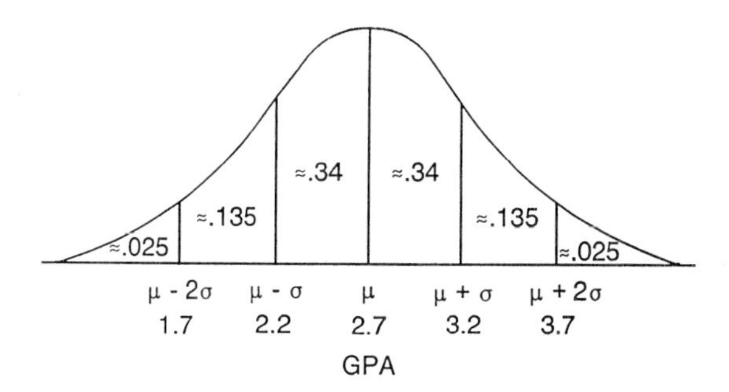

From the Empirical Rule, we know that ≈.025 or ≈2.5% of the students will have GPAs above 3.7 (with $z = 2$). Thus, the GPA corresponding to summa cum laude (top 2.5%) will be greater than 3.7 ($z > 2$).

We know that ≈.16 or 16% of the students will have GPAs above 3.2 ($z = 1$). Thus, the limit on GPAs for cum laude (top 16%) will be greater than 3.2 ($z > 1$).

We must assume the distribution is mound shaped.

2.81 a. If the distribution of scores was symmetric, the mean and median would be equal. The fact that the mean exceeds the median is an indication that the distribution of scores is skewed to the right.

b. It means that 90% of the scores are below 660, and 10% are above 660. (This ignores the possibility of ties, i.e., other people obtaining a score of 660.)

c. If you scored at the 94th percentile, 94% of the scores are below your score, while 6% exceed your score.

2.83 a. $\bar{x} = \frac{\sum_{i=1}^{n} x_i}{n} = \frac{72.4}{9} = 8.04$

$$s = \sqrt{\frac{\sum_{i=1}^{n} x_i^2 - \frac{\left(\sum_{i=1}^{n} x_i\right)^2}{n}}{n - 1}} = \sqrt{\frac{690.64 - \frac{72.4^2}{9}}{9 - 1}} = \sqrt{13.5278} = 3.68$$

b. United States: $z = \frac{x - \bar{x}}{s} = \frac{9.6 - 8.04}{3.68} = .42$

Australia: $z = \frac{x - \bar{x}}{s} = \frac{10.0 - 8.04}{3.68} = .53$

Japan: $z = \frac{x - \bar{x}}{s} = \frac{2.7 - 8.04}{3.68} = -1.45$

c. Since the z-scores for the United States and Australia are positive, the unemployment rates of these 2 countries are above the average unemployment rate. The z-score for Japan was negative, indicating that the unemployment rate in Japan is below the average unemployment rate.

2.85 We first evaluate how extreme a value of 200 ppm for a PCB count is by converting this amount to a z-score

$$z = \frac{200 - 25}{5} = \frac{175}{5} = 35$$

By using Chebyshev's rule, the proportion of times a PCB count of 200 should occur is at most

$$1/35^2 = 1/1225 \approx .0008$$

With such a small probability for the maximum probability of observing this value, it would definitely be considered to be an outlier.

2.87 The interquartile range is IQR = $Q_U - Q_L = 85 - 60 = 25$.

The lower inner fence = $Q_L - 1.5(IQR) = 60 - 1.5(25) = 22.5$.

The upper inner fence = $Q_U + 1.5(IQR) = 85 + 1.5(25) = 122.5$.

The lower outer fence = $Q_L - 3(IQR) = 60 - 3(25) = -15$.

The upper outer fence = $Q_U + 3(IQR) = 85 + 3(25) = 160$.

With only this information, the box plot would look something like the following:

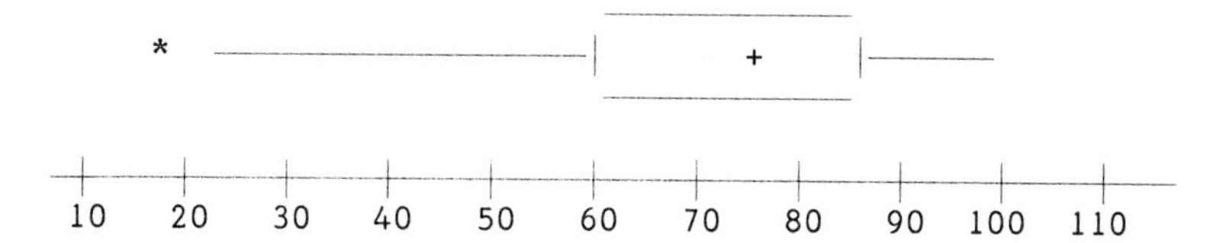

The whiskers extend to the inner fences unless no data points are that small or that large. The upper inner fence is 122.5. However, the largest data point is 100, so the whisker stops at 100. The lower inner fence is 22.5. The smallest data point is 18, so the whisker extends to 22.5. Since 18 is between the inner and outer fences, it is designated with a *. We do not know if there is any more than one data point below 22.5, so we cannot be sure that the box plot is entirely correct.

2.89 For the first data set, the median appears to be about 3.6, and the interquartile range is 5.6 - 2.8 = 2.8. Thus, the middle 75% of the data ranges only 2.8 units. For data set 2, the median appears to be about -2 and the interquartile range is 10 - (-10) = 20. Thus, the middle 75% of the data ranges 20 units. Data set 2 is more variable than data set 1 when comparing the middle 50% of the data.

The first data set appears to be skewed to the right because the median is to the left of the center of the box; the whisker to the right is longer than the one to the left; one observation lies outside the inner fence (about 14.8); and two observations lie outside the outer fence (about 18.8 and 19.6). The second data set appears to be more symmetrical. The median falls near the center of the box; the two whiskers are about the same length, while no observations lie outside the inner fences.

Since the first data set had two observations that lie outside the outer fence, they appear to be outliers. There is one additional suspect outlier which lies between the inner and outer fence. The second data set has no outliers.

2.91 a. Using Minitab, the box plot is:

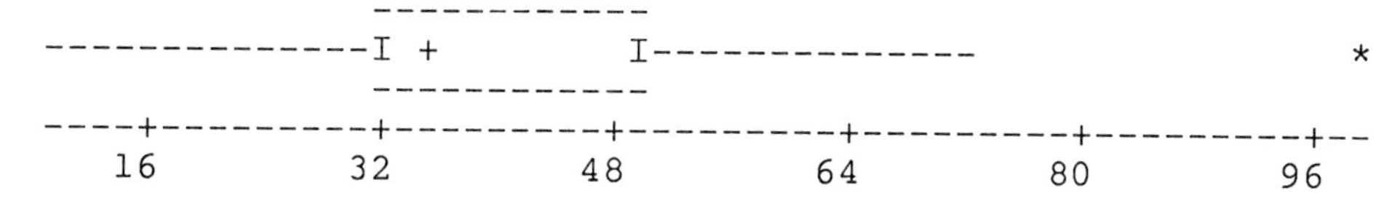

b. There is one suspect outlier, 99. It lies between the inner and outer fences.

2.93 a. The median is about 1540, while the upper quartile is about 3078 and the lower quartile is about 770.

b. The distribution appears to be skewed because the upper whisker is longer than the lower whisker and there are data points above the upper whisker but not below the lower whisker.

c. There are 2 outliers above the upper outer fence and one suspect outlier between the inner and outer fences. The 2 outliers are about 9234 and 12312 while the suspect outlier is about 6800.

2.95 a. Using Minitab, the box plot is:

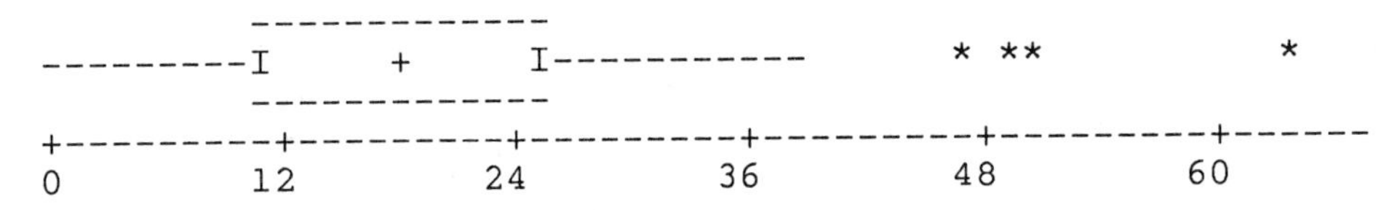

The median is about 18. The data appear to be skewed to the right since there are 4 suspect outliers to the right and none to the left. The variability of the data is fairly small because the IQR is fairly small, approximately 26 - 10 = 16.

b. The customers associated with the suspected outliers are customers 238, 268, 269, and 264.

c. In order to find the z-scores, we must first find the mean and standard deviation.

$$\bar{x} = \frac{\sum x}{n} = \frac{815}{40} = 20.375$$

$$s^2 = \frac{1}{n-1}\{\sum x^2 - \frac{(\sum x)^2}{n}\} = \frac{1}{39}\{24129 - \frac{815^2}{40}\} = 192.90705$$

$$s = \sqrt{192.90705} = 13.89$$

The z-scores associated with the suspected outliers are:

Customer 238 $z = \frac{x - \bar{x}}{s} = \frac{47 - 20.375}{13.89} = 1.92$

Customer 268 $z = \frac{49 - 20.375}{13.89} = 2.06$

Customer 269 $z = \frac{50 - 20.375}{13.89} = 2.13$

Customer 264 $z = \frac{64 - 20.375}{13.89} = 3.14$

2.97 a. Using Minitab, the box plots are:

1960

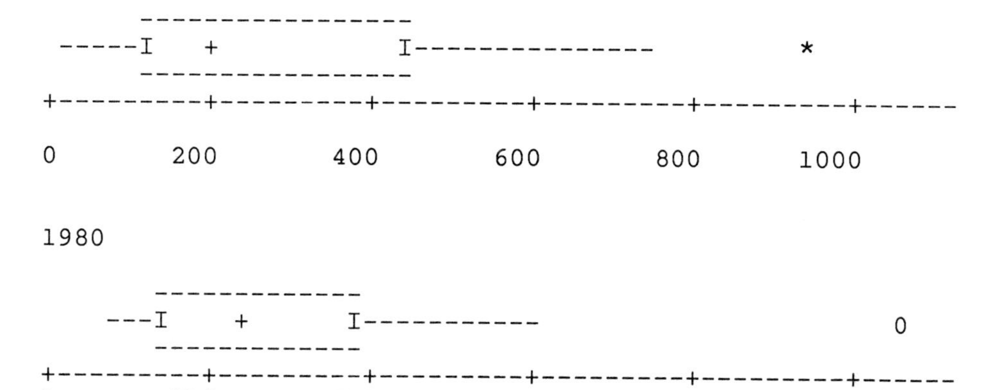

b. The median for 1960 is approximately 200, while the median for 1980 is approximately 360. Thus, the median population size is much larger in 1980 than in 1960.

The IQR for 1960 is approximately 440 - 120 = 320. The IQR for 1980 is approximately 570 - 210 = 360. Since the IQR for the 2 years are very similar, the variability for the 2 years are about the same. Both distributions appear to be skewed to the right. The distribution for 1960 has one suspect outlier (between the inner and outer fences), while the distribution in 1980 has one outlier (lies outside the outer fence).

c. We have no data on nonsunbelt cities. Thus, we cannot compare the growth of sunbelt cities versus nonsunbelt cities.

2.99 a. Using Minitab, the box plot is:

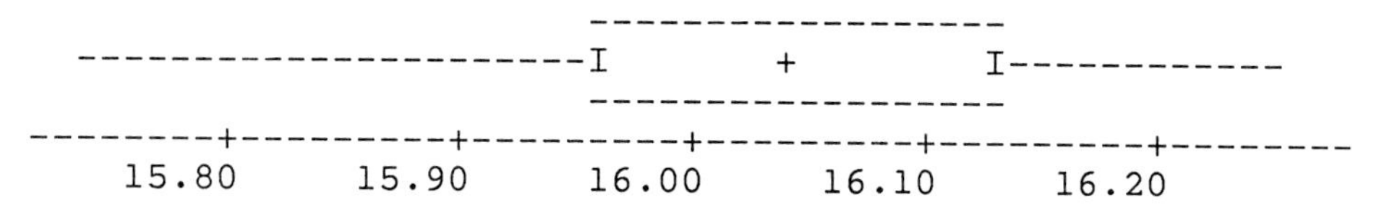

The median disk diameter is about 16.04. The IQR is approximately 16.13 - 15.96 = .17. The distribution is somewhat skewed to the left since the left whisker is longer than the right whisker.

b. Since there are no data points outside the inner fences, there appears to be no outliers.

2.101 The mean is sensitive to extreme values in a data set. Therefore, the median is preferred to the mean when a data set has extreme values present.

2.103 The relative frequency histrogram is:

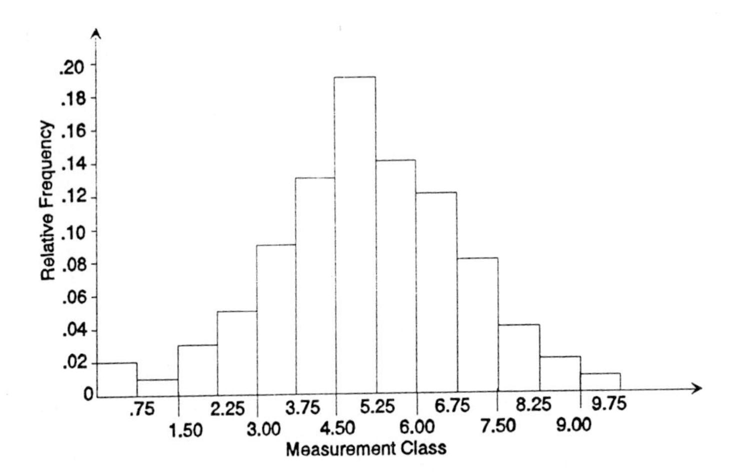

2.105 a. $z = \frac{x - \mu}{\sigma} = \frac{50 - 60}{10} = -1$

$z = \frac{70 - 60}{10} = 1$

$z = \frac{80 - 60}{10} = 2$

b. $z = \frac{x - \mu}{\sigma} = \frac{50 - 60}{5} = -2$

$z = \frac{70 - 60}{5} = 2$

$z = \frac{80 - 60}{5} = 4$

c. $z = \frac{x - \mu}{\sigma} = \frac{50 - 40}{10} = 1$

$z = \frac{70 - 40}{10} = 3$

$z = \frac{80 - 40}{10} = 4$

d. $z = \frac{x - \mu}{\sigma} = \frac{50 - 40}{100} = .1$

$z = \frac{70 - 40}{100} = .3$

$z = \frac{80 - 40}{100} = .4$

2.107 a. $\sum_{i=1}^{n} x_i = 4 + 6 + 6 + 5 + 6 + 7 = 34$

$\sum_{i=1}^{n} x_i^2 = 4^2 + 6^2 + 6^2 + 5^2 + 6^2 + 7^2 = 198$

$$\bar{x} = \frac{\sum_{i=1}^{n} x_i}{n} = \frac{34}{6} = 5.67$$

$$s^2 = \frac{\sum_{i=1}^{n} x_i^2 - \frac{\left(\sum_{i=1}^{n} x_i\right)^2}{n}}{n - 1} = \frac{198 - \frac{34^2}{6}}{6 - 1} = \frac{5.3333}{5} = 1.067$$

$s = \sqrt{1.067} = 1.03$

b. $\sum_{i=1}^{n} x_i = -1 + 4 + (-3) + 0 + (-3) + (-6) = -9$

$\sum_{i=1}^{n} x_i^2 = (-1)^2 + 4^2 + (-3)^2 + 0^2 + (-3)^2 + (-6)^2 = 71$

$$\bar{x} = \frac{\sum_{i=1}^{n} x_i}{n} = \frac{-9}{6} = -\$1.5$$

$$s^2 = \frac{\sum_{i=1}^{n} x_i^2 - \frac{\left(\sum_{i=1}^{n} x_i\right)^2}{n}}{n - 1} = \frac{71 - \frac{(-9)^2}{6}}{6 - 1} = \frac{57.5}{5} = 11.5 \text{ dollars squared}$$

$s = \sqrt{11.5} = \$3.39$

c. $\sum_{i=1}^{n} x_i = \frac{3}{5} + \frac{4}{5} + \frac{2}{5} + \frac{1}{5} + \frac{1}{16} = 2.0625$

$$\sum_{i=1}^{n} x_i^2 = \left(\frac{3}{5}\right)^2 + \left(\frac{4}{5}\right)^2 + \left(\frac{2}{5}\right)^2 + \left(\frac{1}{5}\right)^2 + \left(\frac{1}{16}\right)^2 = 1.2039$$

$$\bar{x} = \frac{\sum_{i=1}^{n} x_i}{n} = \frac{2.0625}{5} = .4125\%$$

$$s^2 = \frac{\sum_{i=1}^{n} x_i^2 - \frac{\left(\sum_{i=1}^{n} x_i\right)^2}{n}}{n - 1} = \frac{1.2039 - \frac{2.0625^2}{5}}{5 - 1} = \frac{.3531}{4} = .088\% \text{ squared}$$

$$s = \sqrt{.088} = .30\%$$

d. (a) Range = 7 - 4 = 3

(b) Range = \$4 - (\$-6) = \$10

(c) Range = $\frac{4}{5}\% - \frac{1}{16}\% = \frac{64}{80}\% - \frac{5}{80}\% = \frac{59}{80}\% = .7375\%$

2.109 The range is found by taking the largest measurement in the data set and subtracting the smallest measurement. Therefore, it only uses two measurements from the whole data set. The standard deviation uses every measurement in the data set. Therefore, it takes every measurement into account--not just two. The range is affected by extreme values more than the standard deviation.

2.111 a. Using the digit in the 10's column as the stem and the digit to the right of the 10's column as the leaf, the stem and leaf display is:

Stem	Leaf
2	8 9 9 7 7 6 8 7 9 8 7
3	0 5 0 0 1 1 2 3 2 1 3 0 0 1 1 1 0 4 0

Because there are only 2 stems, not much can be said about the distribution. There are more observations 30 and above than below 30.

b. To construct a relative frequency histogram, we first select the number of measurement classes. Suppose 6 (arbitrarily) is chosen. The class width is:

$$\text{Class width} = \frac{\text{Largest - Smallest measurement}}{\text{Number of intervals}}$$

$$= \frac{35 - 26}{6} = 1.5 \approx 2$$

The lower class interval will begin at .5 below the smallest measurement or 25.5. The frequency distribution is:

MEASUREMENT CLASS	FREQUENCY	RELATIVE FREQUENCY
25.5-27.5	5	5/30 = .167
27.5-29.5	6	6/30 = .200
29.5-31.5	13	13/30 = .433
31.5-33.5	4	4/30 = .133
33.5-35.5	2	2/30 = .067
	n = 30	1.000

The relative frequency histogram is:

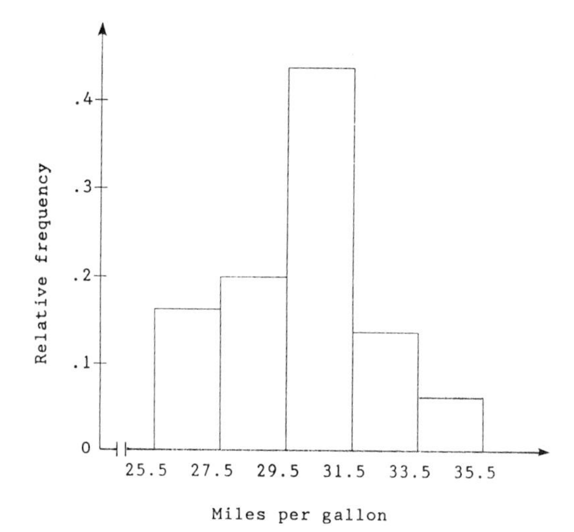

c. $\bar{x} = \frac{\sum x}{n} = \frac{900}{30} = 30$

$$s^2 = \frac{\sum x^2 - \frac{(\sum x)^2}{n}}{n - 1} = \frac{27140 - \frac{900^2}{30}}{30 - 1} = \frac{140}{29} = 4.8276$$

$s = \sqrt{4.8276} = 2.1972$

d. Since the sample mean is 30, this is a good estimate of the population mean. The manufacturer should be satisfied.

e. By Chebyshev's rule,

at least 0% should fall within $\bar{x} \pm s$,
at least 3/4 should fall within $\bar{x} \pm 2s$,
and at least 8/9 should fall within $\bar{x} \pm 3s$.

f. $\bar{x} \pm s \Rightarrow 30 \pm 2.1972 \Rightarrow (27.8028, 32.1972)$. There are actually 21 observations or 21/30 x 100% = 70% in the interval.

$\bar{x} \pm 2s \Rightarrow 2(2.1972) \Rightarrow (25.6056, 34.3944)$. There are actually 29 observations or 29/30 x 100% = 96.7% in the interval.

$\bar{x} \pm 3s \Rightarrow 30 \pm 3(2.1972) \Rightarrow (23.4084, 36.5916)$. There are actually 30 observations or 100% in the interval.

These percentages are very close to the percentages given in part (e).

2.113 a. The stem consists of the digits in the "tens" column, and the leaf consists of the digits to the right of the "tens" column. The stem and leaf display is

Stem	Leaf
4	4 9 8 6
5	4 7 1 8 5 2 9 8 0
6	0 3 4 2 0 9 7 5 5 2
7	5 3

Most of the observations fall between 50 and 70. There do not appear to be any outliers.

b. To construct a relative frequency histogram, we first select the number of measurement classes. Suppose 6 (arbitrarily) is chosen. The class width is:

$$\text{Class width} = \frac{\text{Largest - Smallest measurement}}{6}$$

$$= \frac{75 - 44}{6} = \frac{31}{6} = 5.167 \approx 6.0$$

The lower measurement class begins .5 below the smallest measurement or 43.5. The frequency distribution is:

MEASUREMENT CLASS	FREQUENCY	RELATIVE FREQUENCY
43.5-49.5	4	4/25 = .160
49.5-55.5	5	5/25 = .200
55.5-61.5	6	6/25 = .240
61.5-67.5	7	7/25 = .280
67.5-73.5	2	2/25 = .080
73.5-79.5	1	1/25 = .040
	n = 25	1.000

The relative frequency histogram is:

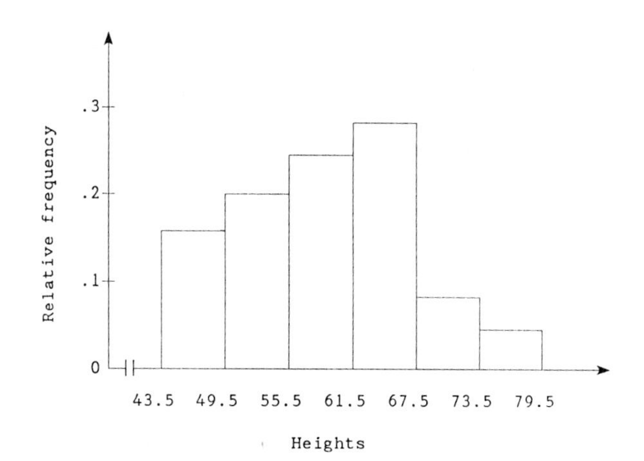

c. $\bar{x} = \frac{\sum x}{n} = \frac{1456}{25} = 58.24$

$$s^2 = \frac{\sum x^2 - \frac{(\sum x)^2}{n}}{n - 1} = \frac{86368 - \frac{1456^2}{25}}{25 - 1} = \frac{1570.56}{24} = 65.44$$

$s = \sqrt{65.44} = 8.0895$

d. Because the data is fairly mound-shaped, we would expect the Empirical Rule to hold. From the Empirical Rule, about 68% of the observations will fall within $\bar{x} \pm s$, about 95% of the observations will fall within $\bar{x} \pm 2s$, and about all will fall within $\bar{x} \pm 3s$.

e. $\bar{x} \pm s$ => 58.24 ± 8.0895 => (50.1505, 66.3295). There are 16 or 16/25 x 100% = 64% in this interval.

$\bar{x} \pm 2s$ => 58.24 ± 2(8.0895) => (42.0610, 74.4190). There are 24 or 24/25 x 100% = 96% in this interval.

$\bar{x} \pm 3s$ => 58.24 ± 3(8.0895) => (33.9715, 82.5085). There are 25 or 100% in this interval.

These percentages are very close to the percentages of the Empirical Rule.

2.115 a. The stem consists of the digits to the left of the "ones" column while the leaf consists of the digit in the "ones" column.

b. In 1982, the smallest score is 790 corresponding to South Carolina, while the highest score is 925 corresponding to New Hampshire. In 1985, the smallest score is 815 corresponding to South Carolina, while the largest score is 939 corresponding to New Hampshire.

c. There are n = 22 observations in each data set. The median is defined as the average of the middle two numbers once they have been arranged in order. For 1982, the median is the average of the 11th and 12th observations which is (888 + 888)/2 = 888. For 1985, the median is the average of the 11th and 12 observations which is (895 + 898)/2 = 896.5

d. Both distributions appear to be skewed to the left, since most observations are large with a few small ones. The scores for 1985 tend to be somewhat larger than those in 1982 because the median for 1985 is 896.5, while the median for 1982 is 888.

e. The 11 stems in Exercise 2.115 appear to give more information than the 4 or 3 stems in Exercise 2.114. The 11 stems allow for more information to be displayed.

2.117 a. First, we must find the z-score associated with each value.

$$z = \frac{x - \mu}{\sigma} = \frac{73 - 83}{10} = -1$$

$$z = \frac{93 - 83}{10} = 1$$

Thus, 73 to 93 is the interval from one standard deviation below to one standard deviation above the mean. The Empirical Rule says approximately 68% of the days will fall between 73 and 93.

b. $z = \dfrac{63 - 83}{10} = -2$ $\qquad$ $z = \dfrac{83 - 83}{10} = 0$

The Empirical Rule says approximately 95% of the days will fall between two standard deviations below and above the mean. The interval 63 to 83 is from two standard deviations below the mean to the mean. Since mound-shaped distributions are symmetric, about 1/2 x 95% or 47.5% of the days will fall between 63 and 83.

c. $z = \frac{93 - 83}{10} = 1$

The Empirical Rule says approximately 68% of the observations will fall between one standard deviation below and above the mean. Since the mound-shaped distribution is symmetric, 1/2 x 68% or 34% of the days will fall between the mean 83 and 93. Since half or 50% of the days fall above the mean and 34% of the days fall between the mean and 93, then 50% - 34% = 16% of the days fall above 93.

2.119 a. First, compute the z-score for 16.

$$z = \frac{x - \mu}{\sigma} = \frac{16 - 16.09}{.03} = -3$$

Using the Empirical Rule, about all the observations fall within 3 standard deviations of the mean. Thus, essentially none of the boxes will contain less than 16 ounces if $\mu = 16.09$.

b. It is very unlikely a box will contain less than 16 ounces if the mean is 16.09 ounces.

c. $z = \frac{x - \mu}{\sigma} = \frac{16.05 - 16.09}{.03} = -1.33$. This is not an unlikely value to obtain, so it is not unlikely to observe a box with 16.05 ounces.

2.121 a. Using Chebyshev's rule, with $\mu = 60$ and $\sigma = 4.5$, we are interested in the percentage of measurements falling above 69, which is $\mu + 2\sigma$. The percentage in the interval $\mu \pm 2\sigma$ is at least 75%; hence, the percentage outside this interval is at most 25%. Since we have no other knowledge of the distribution of scores (that is, no symmetry or other information), we can say only that at most 25% lie in this interval.

b. Using the Empirical Rule, approximately 95% lie in the interval $\mu \pm 2\sigma$; approximately 5% are above 69 or below 51, and by symmetry, approximately (5)% = 2.5% are above 69.

CHAPTER

3

PROBABILITY

3.1 a. Since the probabilities must sum to 1,

$$P(E_3) = 1 - P(E_1) - P(E_2) - P(E_4) - P(E_5)$$
$$= 1 - .1 - .3 - .1 - .1 = .4$$

b. $P(E_3) = 1 - P(E_3) - P(E_2) - P(E_4) - P(E_5)$

$$\Rightarrow 2P(E_3) = 1 - .1 - .2 - .2$$
$$\Rightarrow 2P(E_3) = .5$$
$$\Rightarrow P(E_3) = .25$$

c. $P(E_3) = 1 - P(E_1) - P(E_2) - P(E_4) - P(E_5)$

$$= 1 - .1 - .1 - .1 - .1 = .6$$

3.3 $P(A) = P(1) + P(2) + P(3) = .05 + .25 + .30 = .60$
$P(B) = P(1) + P(3) + P(5) = .05 + .30 + .15 = .50$
$P(C) = P(1) + P(2) + P(3) + P(5) = .05 + .25 + .30 + .15 = .75$

3.5 Each student will obtain slightly different relative frequencies. However, the relative frequencies should be close to $P(A) = 7/8$, $P(B) = 3/8$, and $P(C) = 1/2$.

3.7 a. $P(A) = P(2) + P(3) + P(5) = 2/9 + 1/9 + 1/9 = 4/9$

b. $P(B) = P(4) + P(6) = 2/9 + 1/9 + 3/9 = 1/3$

c. Since there are no events that occur in both A and B at the same time, P(A and B simultaneously) = 0.

3.9 The simple events are:

1, H	1, T
2, H	2, T
3, H	3, T
4, H	4, T
5, H	5, T
6, H	6, T

Since each event is equally likely if the die and coin are fair, each event will have a probability of 1/12.

P(A) = P(6, H) = 1/12

P(B) = P(2, T) + P(4, T) + P(6, T) = 1/12 + 1/12 + 1/12 = 3/12 = 1/4

P(C) = P(2, T) + P(4, T) + P(6, T) + P(2, H) + P(4, H) + P(6, H)
= 1/12 + 1/12 + 1/12 + 1/12 + 1/12 + 1/12 = 6/12 = 1/2

P(D) = P(1, T) + P(2, T) + P(3, T) + P(4, T) + P(5, T) + P(6, T)
= 1/12 + 1/12 + 1/12 + 1/12 + 1/12 + 1/12 = 6/12 = 1/2

3.11 Each student will obtain slightly different proportions. However, the proportions should be close to P(A) = 1/10, P(B) = 6/10 and P(C) = 3/10.

3.13 The simple events are:

W, W
W, B
B, W
B, B

Since the probability that a child born of a mixed marriage is white is 1/2 and the probability the child is black is 1/2, each of the simple events is equally likely. Thus, each simple event has probability 1/4.

a. P(both black) = P(B, B) = 1/4

b. P(1 white, 1 black) = P(W, B) + P(B, W) = 1/4 + 1/4 = 1/2

c. P(both white) = P(W, W) = 1/4

3.15 a. The simple events are:

VM where person chosen has Visa or Mastercard
DC where person chosen has Diners Club
CB where person chosen has Carte Blanche
C where person chosen has Choice

b. The probabilities of the simple events correspond to their relative frequencies. The total sample size is n = 6.0 + 2.2 + .3 + 1 = 9.5.

P(VM) = 6/9.5 = .632
P(DC) = 2.2/9.5 = .232
P(CB) = .3/9.5 = .032
P(C) = 1/9.5 = .105

c. P(person uses one of Citicorp's own credit cards)
= P(DC) + P(CB) + P(C) = .232 + .032 + .105 = .369

3.17 Let D_1 and D_2 correspond to the 2 cars that have had their odometers set back and N_1 and N_2 correspond to the 2 cars that have not had their odometers set back. The sample space is:

D_1D_2	D_2N_1
D_1N_1	D_2N_2
D_1N_2	N_1N_2

Assuming each event is equally likely, each has a probability of 1/6.

a. P(both set back) = $P(D_1D_2)$ = 1/6

b. P(both untampered) = $P(N_1N_2)$ = 1/6

c. P(exactly 1 has odometer set back)
$= P(D_1N_1) + P(D_1N_2) + P(D_2N_1) + P(D_2N_2)$
$= 1/6 + 1/6 + 1/6 + 1/6 = 4/6 = 2/3$

3.19 If each of 3 players uses a fair coin, the sample space is:

HHH	HTT
HHT	THT
HTH	TTH
THH	TTT

Since each event is equally likely, each event will have probability 1/8.

The probability of odd man out on the first roll is

P(HHT) + P(HTH) + P(THH) + P(HTT) + P(THT) + P(TTH)
= 1/8 + 1/8 + 1/8 + 1/8 + 1/8 + 1/8 = 6/8 = 3/4

If one player uses a two-headed coin, the sample space will be (assume 3rd player has the two headed coin):

HHH	THH
HTH	TTH

Since each event is equally likely, each will have probability 1/4.

The probability the 3rd player is odd man out is

P(TTH) = 1/4

From the first part, the probability the 3rd player is odd man out is P(HHT) + P(TTH) = 1/8 + 1/8 = 1/4. Thus, the two probabilities are the same.

3.21 Let P_1 and P_2 correspond to the two new issues that show a profit and N_1, N_2 and N_3 correspond to the three new issues that do not show a profit. The sample space is:

P_1P_2	P_2N_1	N_1N_3
P_1N_1	P_2N_2	N_2N_3
P_1N_2	P_2N_3	
P_1N_3	N_1N_2	

Assuming each event is equally likely, each will have probability 1/10.

a. $P(\text{both show profit}) = P(P_1P_2) = 1/10$

b. $P(\text{neither show profit}) = P(N_1N_2) + P(N_1N_3) + P(N_2N_3)$
$= 1/10 + 1/10 + 1/10 = 3/10$

c. $P(\text{at least one shows profit})$
$= P(P_1P_2) + P(P_1N_1) + P(P_1N_2) + P(P_1N_3) + P(P_2N_1) + P(P_2N_2) + P(P_2N_3)$
$= 1/10 + 1/10 + 1/10 + 1/10 + 1/10 + 1/10 + 1/10 = 7/10$

3.23 a. The odds in favor of Snow Chief are $\frac{1}{3}$ to $(1 - \frac{1}{3})$ or $\frac{1}{3}$ to $\frac{2}{3}$ or 1 to 2.

b. If the odds are 1 to 1, $P(\text{Snow Chief will win}) = \frac{1}{1 + 1} = \frac{1}{2} = .5$

c. If the odds against Snow Chief winning are 3 to 2, the odds for Snow Chief winning are 2 to 3. The probability Snow Chief will win is $\frac{2}{2 + 3} = \frac{2}{5} = .4$

3.25 The experiment consists of rolling a pair of fair dice. The simple events are:

1, 1	2, 1	3, 1	4, 1	5, 1	6, 1
1, 2	2, 2	3, 2	4, 2	5, 2	6, 2
1, 3	2, 3	3, 3	4, 3	5, 3	6, 3
1, 4	2, 4	3, 4	4, 4	5, 4	6, 4
1, 5	2, 5	3, 5	4, 5	5, 5	6, 5
1, 6	2, 6	3, 6	4, 6	5, 6	6, 6

Since each die is fair, each simple event is equally likely. The probability of each simple event is 1/36.

a. A: $\{(1, 6), (2, 5), (3, 4), (4, 3), (5, 2), (6, 1)\}$

B: $\{(1, 4), (2, 4), (3, 4), (4, 4), (5, 4), (6, 4), (4, 1), (4, 2), (4, 3), (4, 5), (4, 6)\}$

$A \cap B$: $\{(3, 4), (4, 3)\}$

$A \cup B$: $\{(1, 4), (2, 4), (3, 4), (4, 4), (5, 4), (6, 4), (4, 1), (4, 2), (4, 3), (4, 5), (4, 6), (1, 6), (2, 5), (5, 2), (6, 1)\}$

A': $\{(1, 1), (1, 2), (1, 3), (1, 4), (1, 5), (2, 1), (2, 2), (2, 3), (2, 4), (2, 6), (3, 1), (3, 2), (3, 3), (3, 5), (3, 6), (4, 1), (4, 2), (4, 4), (4, 5), (4, 6), (5, 1), (5, 3), (5, 4), (5, 5), (5, 6), (6, 2), (6, 3), (6, 4), (6, 5), (6, 6)\}$

b. $P(A) = 6\left(\frac{1}{36}\right) = \frac{6}{36} = \frac{1}{6}$

$P(B) = 11\left(\frac{1}{36}\right) = \frac{11}{36}$

$P(A \cap B) = 2\left(\frac{1}{36}\right) = \frac{2}{36} = \frac{1}{18}$

$P(A \cup B) = 15\left(\frac{1}{36}\right) = \frac{15}{36} = \frac{5}{12}$

$P(A') = 30\left(\frac{1}{36}\right) = \frac{30}{36} = \frac{5}{6}$

3.27 a. $P(A') = P(E_3) + P(E_6) = .2 + .3 = .5$

b. $P(B') = P(E_1) + P(E_7) = .13 + .06 = .19$

c. $P(A' \cap B) = P(E_3) + P(E_6) = .2 + .3 = .5$

d. $P(A \cup B) = P(E_1) + P(E_2) + P(E_3) + P(E_4) + P(E_5) + P(E_6) + P(E_7)$
$= .13 + .05 + .2 + .2 + .06 + .3 + .06 = 1.00$

e. $P(A \cap B) = P(E_2) + P(E_4) + P(E_5) = .05 + .20 + .06 = .31$

f. $P(A' \cup B') = P(E_1) + P(E_7) + P(E_3) + P(E_6)$
$= .13 + .06 + .20 + .30 = .69$

3.29 a. The event is $A \cap F$.

b. The event as the union of 2 events is $B \cup C$.
The event as the complement of an event is A'.

3.31 a. The event is $B \cap C$.

b. The event is A'.

c. The event is $C \cup B$.

d. The event is $A \cap C'$.

3.33 a. The event $A \cap B$ is the event the outcome is black and odd. The event is $A \cap B$: {11, 13, 15, 17, 29, 31, 33, 35}

b. The event $A \cup B$ is the event the outcome is black or odd or both. The event $A \cup B$ is {2, 4, 6, 8, 10, 11, 13, 15, 17, 20, 22, 24, 26, 28, 29, 31, 33, 35, 1, 3, 5, 7, 9, 19, 21, 23, 25, 27}

c. Assuming all events are equally likely, each has a probability of 1/38.

$$P(A) = 18\left(\frac{1}{38}\right) = \frac{18}{38} = \frac{9}{19}$$

$$P(B) = 18\left(\frac{1}{38}\right) = \frac{18}{38} = \frac{9}{19}$$

$$P(A \cap B) = 8\left(\frac{1}{38}\right) = \frac{8}{38} = \frac{4}{19}$$

$$P(A \cup B) = 28\left(\frac{1}{38}\right) = \frac{28}{38} = \frac{14}{19}$$

$$P(C) = 18\left(\frac{1}{38}\right) = \frac{18}{38} = \frac{9}{19}$$

d. The event $A \cap B \cap C$ is the event the outcome is odd and black and low. The event $A \cap B \cap C$ is {11, 13, 15, 17}.

e. $P(A \cap B \cap C) = 4\left(\frac{1}{38}\right) = \frac{4}{38} = \frac{2}{19}$

f. The event $A \cup B \cup C$ is the event the outcome is odd or black or low. The event $A \cup B \cup C$ is

{1, 2, 3, ... , 29, 31, 33, 35}

or

{all simple events except 00, 0, 30, 32, 34, 36}

g. $P(A \cup B \cup C) = 32\left(\frac{1}{38}\right) = \frac{32}{38} = \frac{16}{19}$

3.35 a. The event $A \cap F$ is the event the shaft selected is from warehouse 1 and it is extra stiff.

$$P(A \cap F) = \frac{3\%}{100\%} = .03$$

b. The event $C \cup E$ is the event the shaft selected is from warehouse 3 or is stiff or both.

$$P(C \cup E) = \frac{28\%}{100\%} + \frac{18\%}{100\%} + \frac{0\%}{100\%} + \frac{8\%}{100\%} + \frac{8\%}{100\%} = \frac{62\%}{100\%} = .62$$

c. The event $C \cap D$ is the event the shaft selected is from warehouse 3 and is regular.

$$P(C \cap D) = \frac{28\%}{100\%} = .28$$

d. The event $A \cup F$ is the event the shaft selected is from warehouse 1 or is extra stiff or both.

$$P(A \cup F) = \frac{19\%}{100\%} + \frac{8\%}{100\%} + \frac{3\%}{100\%} + \frac{2\%}{100\%} + \frac{0\%}{100\%} = \frac{32\%}{100\%} = .32$$

e. The event $A \cup D$ is the event the shaft selected is from warehouse 1 or is regular or both.

$$P(A \cup D) = \frac{19\%}{100\%} + \frac{8\%}{100\%} + \frac{3\%}{100\%} + \frac{14\%}{100\%} + \frac{28\%}{100\%} = \frac{72\%}{100\%} = .72$$

3.37 a. $P(A) = P(E_1) + P(E_2) + P(E_3) = .1 + .1 + .3 = .5$
$P(B) = P(E_2) + P(E_3) + P(E_5) = .1 + .3 + .1 = .5$
$P(A \cap B) = P(E_2) + P(E_3) = .1 + .3 = .4$

b. $$P(E_1|A) = \frac{P(E_1 \cap A)}{P(A)} = \frac{.1}{.5} = .2$$

$$P(E_2|A) = \frac{P(E_2 \cap A)}{P(A)} = \frac{.1}{.5} = .2$$

$$P(E_3|A) = \frac{P(E_3 \cap A)}{P(A)} = \frac{.3}{.5} = .6$$

We are given that $P(E_1) = .1$, $P(E_2) = .1$, and $P(E_3) = .3$

Thus, $P(E_1) = P(E_2)$ and $P(E_3) = 3P(E_1)$

From above, $P(E_1|A) = P(E_2|A)$ and $P(E_3|A) = 3P(E_1|A)$

c. Using the sum of the conditional probabilities,

$$P(B|A) = P(E_2|A) + P(E_3|A) = .2 + .6 = .8$$

Using the formula,

$$P(B|A) = \frac{P(A \cap B)}{P(A)} = \frac{.4}{.5} = .8$$

3.39 a. $P(A) = P(1) + P(2) + P(3) = .20 + .05 + .30 = .55$
$P(B) = P(3) + P(4) = .30 + .10 = .40$
$P(C) = P(5) + P(6) = .10 + .25 = .35$

b. $P(A \cap B) = P(3) = .30$
$P(A \cap C) = 0$
$P(B \cap C) = 0$

c. $$P(1|A) = \frac{P(1 \cap A)}{P(A)} = \frac{.20}{.55} = \frac{4}{11}$$

$$P(2|A) = \frac{P(2 \cap A)}{P(A)} = \frac{.05}{.55} = \frac{1}{11}$$

$$P(3|A) = \frac{P(3 \cap A)}{P(A)} = \frac{.30}{.55} = \frac{6}{11}$$

The sum of these probabilities is $4/11 + 1/11 + 6/11 = 11/11 = 1$

From the given probabilities, $P(1) = 4P(2)$
$P(3) = 6P(2)$

The probabilities in part (c) are in the same proportions.

d. Using the conditional probabilities,

$$P(B|A) = P(3|A) = \frac{6}{11}$$

Using the formula,

$$P(B|A) = \frac{P(A \cap B)}{P(A)} = \frac{.30}{.55} = \frac{6}{11}$$

e. $$P(C|A) = \frac{P(A \cap C)}{P(A)} = \frac{0}{.55} = 0$$

$P(C \cap A') = P(5) + P(6) = .10 + .25 = .35$

$P(A') = 1 - P(A) = 1 - .55 = .45$

$$P(C|A') = \frac{P(C \cap A')}{P(A')} = \frac{.35}{.45} = \frac{7}{9}$$

3.41 Let W_1 and W_2 represent the two white chips, R_1 and R_2 represent the two red chips, and B_1 and B_2 represent the two blue chips. The sample space is:

W_1W_2	W_2R_1	R_1B_1
W_1R_1	W_2R_2	R_1B_2
W_1R_2	W_2B_1	R_2B_1
W_1B_1	W_2B_2	R_2B_2
W_1B_2	R_1R_2	B_1B_2

Assuming each event is equally likely, each event will have a probability of 1/15.

Then $P(A) = P(W_1W_2) + P(R_1R_2) + P(B_1B_2) = 3\left(\frac{1}{15}\right) = \frac{3}{15} = \frac{1}{5}$

$$P(B) = P(R_1R_2) = \frac{1}{15}$$

$$\begin{aligned} P(C) = {} & P(W_1W_2) + P(W_1R_1) + P(W_1R_2) + P(W_1B_1) \\ & + P(W_1B_2) + P(W_2R_1) + P(W_2R_2) + P(W_2B_1) \\ & + P(W_2B_2) + P(R_1R_2) + P(R_1B_1) + P(R_1B_2) \\ & + P(R_2B_1) + P(R_2B_2) \\ = {} & 14\left(\frac{1}{15}\right) = \frac{14}{15} \end{aligned}$$

$$P(A \cap B) = P(R_1R_2) = \frac{1}{15}$$

$$P(A') = 1 - P(A) = 1 - \frac{1}{5} = \frac{4}{5}$$

$$P(A' \cap B) = 0$$

$$P(B \cap C) = P(R_1R_2) = \frac{1}{15}$$

$$P(A \cap C) = P(W_1W_2) + P(R_1R_2) = 2\left(\frac{1}{15}\right) = \frac{2}{15}$$

$$\begin{aligned} P(A' \cap C) = {} & P(W_1R_1) + P(W_1R_2) + P(W_1B_1) + P(W_1B_2) + P(W_2R_1) \\ & + P(W_2R_2) + P(W_2B_1) + P(W_2B_2) + P(R_1B_1) + P(R_1B_2) \\ & + P(R_2B_1) + P(R_2B_2) \\ = {} & 12\left(\frac{1}{15}\right) = \frac{12}{15} = \frac{4}{5} \end{aligned}$$

$$P(B|A) = \frac{P(A \cap B)}{P(A)} = \frac{\frac{1}{15}}{\frac{1}{5}} = \frac{1}{3}$$

$$P(B|A') = \frac{P(A' \cap B)}{P(A')} = \frac{0}{\frac{4}{5}} = 0$$

$$P(B|C) = \frac{P(B \cap C)}{P(C)} = \frac{\frac{1}{15}}{\frac{14}{15}} = \frac{1}{14}$$

$$P(A|C) = \frac{P(A \cap C)}{P(C)} = \frac{\frac{2}{15}}{\frac{14}{15}} = \frac{1}{7}$$

$$P(C|A') = \frac{P(A' \cap C)}{P(A')} = \frac{\frac{4}{5}}{\frac{4}{5}} = 1$$

3.43 Suppose we create a table with the 4 rows corresponding to the 4 income levels and the 2 columns corresponding to the number of filers audited and number of filers not audited. To compute the number of filers in each income level who were audited, multiply the percentage of filers audited by the number of filers divided by 100%. The number not audited is found by subtracting the number audited from the number of filers in each income level. The table is:

(IN MILLIONS)

INCOME	NUMBER OF FILERS AUDITED	NUMBER OF FILERS NOT AUDITED	NUMBER OF FILERS
Under $10,000	.14900	29.65100	29.8
$10,000-$24,999	.28135	32.81865	33.1
$25,000-$49,999	.38080	26.81920	27.2
$50,000 or over	.25536	11.14464	11.4
	1.06651	100.43349	101.5

a. Let A = {taxpayer is audited}

$$P(A) = \frac{1.06651}{101.5} = .0105$$

b. Let B = {taxpayer had income $10,000-$24,999}
C = {taxpayer had income of $50,000 or more}

$$P(A \cap B) = \frac{.28135}{101.5} = .0028$$

$$P(C \cup A') = \frac{.25536}{101.5} + \frac{11.14464}{101.5} + \frac{29.651}{101.5} + \frac{32.81865}{101.5} + \frac{26.81920}{101.5}$$

$$= \frac{100.68885}{101.5} = .9920$$

c. $$P(C|A) = \frac{P(A \cap C)}{P(A)} = \frac{\frac{.25536}{101.5}}{\frac{1.06651}{101.5}} = .2394$$

Let D = {taxpayer had income less than $10,000}

$$P(D|A) = \frac{P(D \cap A)}{P(A)} = \frac{\frac{.14900}{101.5}}{\frac{1.06651}{101.5}} = .1397$$

d. $$P(A|C) = \frac{P(A \cap C)}{P(C)} = \frac{\frac{.25536}{101.5}}{\frac{11.4}{101.5}} = .0224$$

3.45 Define the following events:

A: {patient receives PMI sheet}
B: {patient was hospitalized}

$P(A) = .20$, $P(A \cap B) = .12$

$$P(B|A) = \frac{P(A \cap B)}{P(A)} = \frac{.12}{.20} = .60$$

3.47 Define the following events:

A: {person chosen is black}
B: {person chosen is PGM type 1-1}

a. $$P(A) = \frac{6.7\% + 3.4\% + .4\%}{100\%} = .105$$

b. $$P(B|A) = \frac{P(A \cap B)}{P(A)} = \frac{\frac{6.7\%}{100\%}}{.105} = .638$$

c. $$P(B|A') = \frac{P(A' \cap B)}{P(A')} = \frac{\frac{46.3\%}{100\%}}{1 - .105} = \frac{.463}{.895} = .517$$

3.49 a. $P(A) = P(E_1) + P(E_3) + .22 + .15 = .37$

b. $P(B) = P(E_2) + P(E_3) + P(E_4) = .31 + .15 + .22 = .68$

c. $P(A \cap B) = P(E_3) + .15$

d. $$P(A|B) = \frac{P(A \cap B)}{P(B)} = \frac{.15}{.68} = .221$$

e. $P(B \cap C) = 0$

f. $$P(C|B) = \frac{P(B \cap C)}{P(B)} = \frac{0}{.68} = 0$$

3.51 a. Let A = {person saw ad} and B = {person shopped at x}

$$P(A) = \frac{100}{200} + \frac{25}{200} = \frac{125}{200} = .625$$

b. $$P(A \cap B) = \frac{100}{200} = .5$$

c. $$P(B|A) = \frac{P(A \cap B)}{P(A)} = \frac{.5}{.625} = .8$$

d. $$P(B) = \frac{100}{200} + \frac{25}{200} = \frac{125}{200} = .625$$

e. A and B are independent if $P(B|A) = P(B)$

$P(B|A) = .8$
$P(B) = .625$

Therefore, A and B are not independent.

f. $A' = \{\text{did not see ad}\}$ and $B' = \{\text{did not shop at x}\}$.
A' and B' are mutually exclusive if $P(A' \cap B') = 0$.

From the table, $P(A' \cap B') = \frac{50}{200} = .25$

Therefore, A' and B' are not mutually exclusive.

3.53 a. $P(A \cap C) = 0 \Rightarrow$ A and C are mutually exclusive.
$P(B \cap C) = 0 \Rightarrow$ B and C are mutually exclusive.

b. $P(A) = P(1) + P(2) + P(3) = .20 + .05 + .30 = .55$
$P(B) = P(3) + P(4) = .30 + .10 = .40$
$P(C) = P(5) + P(6) = .10 + .25 = .35$
$P(A \quad B) = P(3) = .30$

$$P(A|B) = \frac{P(A \cap B)}{P(B)} = \frac{.30}{.40} = .75$$

A and B are independent if $P(A|B) = P(A)$. Since $P(A|B) = .75$ and $P(A) = .55$, A and B are not independent.

Since A and C are mutually exclusive, they are not independent. Similarly, since B and C are mutually exclusive, they are not independent.

c. Using the probabilities of simple events,

$$P(A \cup B) = P(1) + P(2) + P(3) + P(4)$$
$$= .20 + .05 + .30 + .10 = .65$$

Using the additive rule,

$$P(A \cup B) = P(A) + P(B) - P(A \cap B)$$
$$= .55 + .40 - .30 = .65$$

Using the probability of simple events,

$$P(A \cup C) = P(1) + P(2) + P(3) + P(5) + P(6)$$
$$= .20 + .05 + .30 + .10 + .25 = .90$$

Using the additive rule,

$$P(A \cup C) = P(A) + P(C) - P(A \cap C)$$
$$= .55 + .35 - 0 = .90$$

3.55 a. $P(A) = \frac{16}{50} + \frac{4}{50} + \frac{4}{50} = \frac{24}{50} = .48$

b. $P(B) = \frac{4}{50} + \frac{6}{50} + \frac{4}{50} + \frac{1}{50} + \frac{1}{50} = \frac{16}{50} = .32$

c. $A \cap B$ = {soft drink is bottled by Coca-Cola but is not a cola}
$P(A \cap B) = P(\text{soft drink is Mr. Pibb or Sprite})$

$$= \frac{4}{50} + \frac{4}{50} = \frac{8}{50} = .16$$

d. $A \cup B$ = {soft drink is bottled by Coca-Cola or is not a cola}
$P(A \cup B) = P(\text{Coca-Cola or Mr. Pibb or Seven-up or Sprite or Nehi Orange or Dr. Pepper})$

$$= \frac{16}{50} + \frac{4}{50} + \frac{6}{50} + \frac{4}{50} + \frac{1}{50} + \frac{1}{50} = \frac{32}{50} = .64$$

e. $P(A|B) = \dfrac{P(A \cap B)}{P(B)} = \dfrac{.16}{.32} = .5$

f. $P(B') = 1 - P(B) = 1 - .32 = .68$

$A \cap B'$ = {soft drink bottled by Coca-Cola and is a cola}
= {Coca-Cola}

$$P(A \cap B') = \frac{16}{50} = .32$$

$$P(A|B') = \frac{P(A \cap B')}{P(B')} = \frac{.32}{.68} = .47$$

3.57 a. The simple events are

$A \cap B$: {microchip lasts through first use and end of first year}
$A \cap B'$: {microchip lasts through first use and fails during first year}
A': {microchip does not last through first use}

b. We are given $P(A') = .10$ and $P(B|A) = .99$

$$P(A \cap B) = P(B|A)P(A) = .99(.9) = .891 \qquad P(A) = 1 - P(A') = 1 - .1 = .9$$

Since the sum of the probabilities associated with all the simple events is 1,

$$P(A \cap B') = 1 - P(A \quad B) - P(A') = 1 - .891 - .10 = .009$$

c. $P(A \cap B) = .891$

3.59 Define the following events.

D: A car is defective (has had its odometer set back).
N: A car is not defective.

Then P(D) = 9/10 = .9 and P(N) = 1 - P(D) = .1. The sample space in this experiment consists of 8 triplets, corresponding to the three cars the dealer has in stock.

1.	DDD	3.	NDD	5.	NND	7.	DNN
2.	DND	4.	DDN	6.	NDN	8.	NNN

We assume that each car's condition (D or N) is independent of any other car's condition.

a. $P(\text{all three defective}) = P(DDD) = P(D \cap D \cap D) = P(D)P(D)P(D) = (.9)^3 = .729$

b. $P(\text{none defective}) = P(NNN) = P(N \cap N \cap N) = P(N)P(N)P(N) = (.1)^3 = .001$

c. $P(\text{exactly one defective}) = P(NND) + P(NDN) + P(DNN)$
$= P(N \cap N \cap D) + P(N \cap D \cap N) + P(D \cap N \cap N)$
$= 3P(N)P(N)P(D) = 3(.1)^2(.9) = .027$

d. P(D) = .9

3.61 a. Refer to the simple events listed in Exercise 3.25.

$$P(\text{win on first roll}) = P(7 \text{ or } 11) = \frac{8}{36} = \frac{2}{9}$$

b. $P(\text{lose on first roll}) = P(\text{sum of } 2) = \frac{1}{36}$

c. If the player rolls a 3 on the first roll, the game will end on the next roll if:
1) the player rolls a 3
2) the player rolls a 7 or an 11
3) the player rolls a 2

This probability is the union of three mutually exclusive events.

$$P(\text{game ends on a second roll}) = \frac{2}{36} + \frac{8}{36} + \frac{1}{36} = \frac{11}{36}$$

3.63 Define the following events.

D: Carrier's child acquires the disease.
N: Carrier's child does not acquire the disease.

A simple event is a triplet consisting of the conditions of the carrier's three children. Note that $P(D) = 1/4$ and $P(N) = 1 - P(D) = 3/4$.

a. $P(NNN) = P(N \cap N \cap N) = P(N)P(N)P(N) = (3/4)^3 = 27/64$, since transmission of the disease to one child is independent of transmission to another.

b. $P(DDD) = P(D \cap D \cap D) = \{P(D)\}^3 = (1/4)^3 = 1/64$

c.
$$\begin{aligned} P(\text{exactly one diseased}) &= P(NND) + P(NDN) + P(DNN) \\ &= P(N \cap N \cap D) + P(N \cap D \cap N) \\ &\quad + P(D \cap N \cap N) \\ &= 3P(N)P(N)P(D) \\ &= 3(1/4)(3/4)^2 = 27/64 \end{aligned}$$

3.65 Define the following events:

A: {antigens match} $P(A) = .001$, $P(A') = 1 - P(A) = 1 - .001 = .999$

From part (b), the event "all three antigens match" is AAA. Assuming the outcomes of each donor are independent,

$$P(AAA) = P(A \cap A \cap A) = P(A)P(A)P(A) = .001^3 = .000000001$$

From part (c), the event "none of the antigens will match" is $A'A'A'$.

$$P(A'A'A') = P(A' \cap A' \cap A') = P(A')P(A')P(A') = .999(.999)(.999) = .997$$

3.67 a. Let the 6 elements be A, B, C, D, E, F.

The number of ways to draw 3 elements is:

ABC	ACE	BCD	BEF
ABD	ACF	BCE	CDE
ABE	ADE	BCF	CDF
ABF	ADF	BDE	CEF
ACD	AEF	BDF	DEF

Using combinatorial mathematics, the number of ways to select a sample of n = 3 elements is

$$\binom{6}{3} = \frac{6!}{3!(6-3)!} = \frac{6 \cdot 5 \cdot 4 \cdot 3 \cdot 2 \cdot 1}{3 \cdot 2 \cdot 1 \cdot 3 \cdot 2 \cdot 1} = 20$$

b. Since there are 20 possibilities, the probability of selecting any particular sample is 1/20.

3.69 First, number the intersections from 1 to 5000. Using the randomnumber table, select a starting point that contains 4 digits. Following either the row or column, select successive 4 digit numbers until 50 different 4 digit numbers between 0001 and 5000 are selected. The intersections corresponding to these 50 four digit numbers are selected for sampling.

The second method requires the rows to be numbered from 00 to 99 and the columns to be numbered from 1 to 50. Using the random number table, select a starting point that contains 2 digits. Following either the row or column, select successive 2 digit numbers in sets of 2. The first 2 digit number will correspond to the row number. The second 2 digit number must be between 1 and 50 and corresponds to the column number. This procedure is followed until 50 pairs of 2 digit numbers are selected that correspond to 50 different intersections.

3.71 First, number the households from 1 to 534,322. Using the random number table, select a starting point that consists of 6 digits. Following either the row or column, select successive 6 digit numbers between 1 and 534,322 until 1000 different 6 digit numbers have been selected. The sample will consist of the 1000 households corresponding to the 1000 different numbers.

3.73 a.

(1, H)	(2, 1)	(3, H)	(4, 1)	(5, H)	(6, 1)
(1, T)	(2, 2)	(3, T)	(4, 2)	(5, T)	(6, 2)
	(2, 3)		(4, 3)		(6, 3)
	(2, 4)		(4, 4)		(6, 4)
	(2, 5)		(4, 5)		(6, 5)
	(2, 6)		(4, 6)		(6, 6)

b. Each simple event is an intersection of two independent events. Each simple event whose first element is 1, 3, or 5 has probability

$$\left(\frac{1}{6}\right)\left(\frac{1}{2}\right) = \frac{1}{12}$$

while each simple event whose first element is 2, 4, or 6 has probability

$$\left(\frac{1}{6}\right)\left(\frac{1}{6}\right) = \frac{1}{36}$$

c. $P(A) = P\{(1, H), (3, H), (5, H)\} = \frac{1}{12} + \frac{1}{12} + \frac{1}{12} = \frac{3}{12} = \frac{1}{4}$

$P(B) = P\{(1, H), (1, T), (3, H), (3, T), (5, H), (5, T)\} = \frac{6}{12} = \frac{1}{2}$

d. A': all except (1, H), (3, H), and (5, H)

B': {(2, 1), (2, 2), (2, 3), (2, 4), (2, 5), (2, 6),
(4, 1), (4, 2), (4, 3), (4, 4), (4, 5), (4, 6),
(6, 1), (6, 2), (6, 3), (6, 4), (6, 5), (6, 6)}

$A \cap B$: {(1, H), (3, H), (5, H)}

$A \cup B$: {(1, H), (1, T), (3, H), (3, T), (5, H), (5, T)}

e. $P(A') = 1 - P(A) = 1 - \frac{1}{4} = \frac{3}{4}$

$P(B') = 1 - P(B) = 1 - \frac{1}{2} = \frac{1}{2}$

$P(A \cap B) = P(A) = \frac{1}{4}$

$P(A \cup B) = P(A) + P(B) - P(A \quad B) = \frac{1}{4} + \frac{1}{2} - \frac{1}{4} = \frac{1}{2}$

$$P(A|B) = \frac{P(A \cap B)}{P(B)} = \frac{1/4}{1/2} = \frac{1}{2}$$

$$P(B|A) = \frac{P(A \cap B)}{P(A)} = \frac{1/4}{1/4} = 1$$

f. $P(A \cap B) \neq 0$, so that A and B are not mutually exclusive.

$P(A|B) \neq P(A)$, so that A and B are not independent.

3.75 a. Because events A and B are independent, we have

$P(A \cap B) = P(A)P(B) = (.3)(.1) = .03$

Thus, $P(A \cap B) \neq 0$, and the two events cannot be mutually exclusive.

b. $P(A|B) = \frac{P(A \cap B)}{P(B)} = \frac{.03}{.1} = .3$ $P(B|A) = \frac{P(A \cap B)}{P(A)} = \frac{.03}{.3} = .1$

c. $P(A \cup B) = P(A) + P(B) - P(A \cap B) = .3 + .1 - .03 = .37$

3.77 a. Define the following event:

F_i: {player makes a foul shot on ith attempt}

$P(F_i) = .8$

$P(F_i') = 1 - P(F_i) = 1 - .8 = .2$

The event "the player scores on both shots" is $F_1 \cap F_2$. If the throws are independent, then

$$P(F_1 \cap F_2) = P(F_1)P(F_2) = .8(.8) = .64$$

The event "the player scores on exactly one shot" is

$$(F_1 \cap F_2') \cup (F_1' \cap F_2)$$

$$\begin{aligned} P(F_1 \cap F_2') \cup (F_1' \cap F_2) &= P(F_1 \cap F_2') + P(F_1' \cap F_2) \\ &= P(F_1)P(F_2') + P(F_1')P(F_2) \\ &= .8(.2) + .2(.8) = .16 + .16 = .32 \end{aligned}$$

The event "the player scores on neither shot" is $F_1' \cap F_2'$.

$$P(F_1' \cap F_2') = P(F_1')P(F_2') = .2(.2) = .04$$

b. We know $P(F_1) = .8$, $P(F_2|F_1) = .9$, and $P(F_2|F_1') = .7$

The probability the player scores on both shots is

$$P(F_1 \cap F_2) = P(F_2|F_1)P(F_1) = .9(.8) = .72$$

The probability the player scores on exactly one shot is

$$\begin{aligned} P(F_1 \cap F_2') + P(F_1' \cap F_2) &= P(F_2'|F_1)P(F_1) + P(F_2|F_1')P(F_1') \\ &= [1 - P(F_2|F_1)]P(F_1) + P(F_2|F_1')P(F_1') \\ &= (1 - .9)(.8) + .7(.2) = .08 + .14 \\ &= .22 \end{aligned}$$

The probability the player scores on neither shot is

$$\begin{aligned} P(F_1' \cap F_2') &= P(F_2'|F_1')P(F_1') = [1 - P(F_2|F_1')]P(F_1') \\ &= (1 - .7)(.2) = .06 \end{aligned}$$

c. Two consecutive foul shots are probably dependent. The outcome of the second shot probably depends on the outcome of the first.

3.79 a. The results of consecutive tosses of a coin are independent. The outcome of one toss probably has no affect on the outcome of the next toss.

b. The opinions of randomly selected individuals are probably independent. The opinion of one individual is not affected by that of another.

c. A major league baseball player's results in two consecutive at-bats are probably dependent. Getting a hit the first time up probably affects what happens the second time up.

d. The amount of gain or loss associated with investments in different stocks bought on the same day and sold on the same day are probably dependent. The stock market has trends. Thus, if one stock is up, the tendency is that another will be up also.

e. The amount of gain or loss associated with investments of different stocks bought and sold at different times are probably independent.

f. The responses of 2 different individuals to the same stimulus are probably independent. The response of one individual will not affect the response of another.

3.81 a.

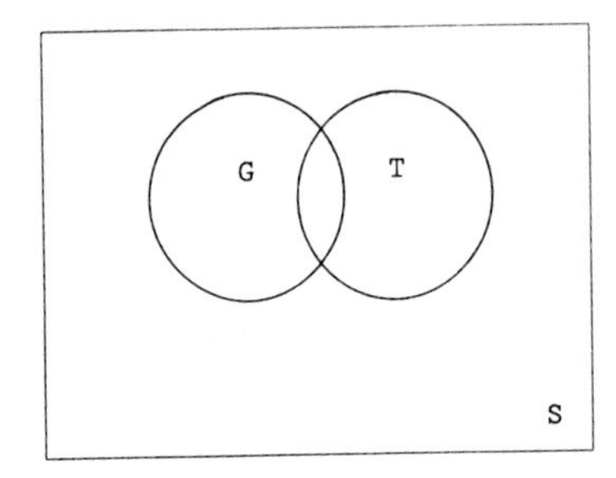

b. Define the following events:

G: {member regularly uses the golf course}
T: {member regularly uses the tennis courts}

$P(G) = .7$, $P(T) = .5$, and $P(G' \cap T') = .05$

The event "the member uses the golf course or the tennis courts or both" is $G \cup T$.

$$P(G \cup T) = 1 - P(G' \cap T') = 1 - .05 = .95$$

c. The event "the member uses both the golf course and the tennis courts" is $G \cap T$.

We know $P(G \cup T) = P(G) + P(T) - P(G \cap T)$

Thus, $P(G \cap T) = P(G) + P(T) - P(G \cup T)$
$= .7 + .5 - .95 = .25$

d. The event "the member uses the golf facilities given he/she uses the tennis courts" is $G|T$.

$$P(G|T) = \frac{P(G \cap T)}{P(T)} = \frac{.25}{.5} = .5$$

3.83 a. Let S denote the event that an insect travels toward the pheromone and F the event that an insect travels toward the control. The sample space associated with the experiment of releasing five insects has 32 simple events:

SSSSS	FSSFS	SSFFF	FFSFS
SSSSF	FSSSF	SFSFF	FFFSS
SSSFS	SFFSS	SFFSF	FFFFS
SSFSS	SFSFS	SFFFS	FFFSF
SFSSS	SFSSF	FSSFF	FFSFF
FSSSS	SSFFS	FSFSF	FSFFF
FFSSS	SSFSF	FSFFS	SFFFF
FSFSS	SSSFF	FFSSF	FFFFF

If the pheromone under study has no effect, then each simple event is equally likely and occurs with probability 1/32.

The probability that all five insects travel toward the pheromone is P(SSSSS) = 1/32.

b. The probability that exactly four of the five insects travel toward the pheromone is

$$P(SSSSF) + P(SSSFS) + P(SSFSS) + P(SFSSS) + P(FSSSS) = 5/32 \approx .156$$

3.85 Define the following events:

C: {committee's judgment is joint is acceptable}
I: {inspector's judgment is joint is acceptable}

The simple event of this experiment are:

$C \cap I$
$C \cap I'$
$C' \cap I$
$C' \cap I'$

a. The probability the inspector judges the joint to be acceptable is:

$$P(I) = P(C \cap I) + P(C' \cap I) = \frac{101}{153} + \frac{23}{153} = \frac{124}{153} \approx .810$$

The probability the committee judges the joint to be acceptable is:

$$P(C) = P(C \cap I) + P(C \cap I') = \frac{101}{153} + \frac{10}{153} = \frac{111}{153} \approx .725$$

b. The probability that both the committee and the inspector judge the joint to be acceptable is:

$$P(C \cap I) = \frac{101}{153} \approx .660$$

The probability that neither judge the joint to be acceptable is:

$$P(C' \cap I') = \frac{19}{153} \approx .124$$

c. The probability the inspector and committee disagree is

$$P(C \cap I') + P(C' \cap I) = \frac{10}{153} + \frac{23}{153} = \frac{33}{153} \approx .216$$

The probability the inspector and committee agree is

$$P(C \cap I) + P(C' \cap I') = \frac{101}{153} + \frac{19}{153} = \frac{120}{153} \approx .784$$

3.87 Define the following events:

S_1: {salesman makes sale on the first visit}
S_2: {salesman makes a sale on the second visit}

$P(S_1) = .4 \qquad P(S_2|S_1') = .65$

The simple events of the experiment are:

$S_1 \cap S_2'$
$S_1' \cap S_2$
$S_1' \cap S_2'$

The probability the salesman will make a sale is

$$\begin{aligned} P(S_1 \cap S_2') + P(S_1' \cap S_2) &= P(S_1) + P(S_2|S_1')P(S_1') \\ &= .4 + .65(1 - .4) \\ &= .4 + .39 = .79 \end{aligned}$$

3.89 a. The number of ways to draw 2 cards from 52 is

$$\binom{52}{2} = \frac{52!}{2!(52-2)!} = \frac{52 \cdot 51 \cdot 50 \cdot \ldots \cdot 1}{2 \cdot 1 \cdot 50 \cdot 49 \cdot \ldots \cdot 1} = 1326$$

In a deck of cards, there are 4 aces and 12 face cards. The number of ways to draw 1 ace and 1 face card is:

$$\begin{aligned} \binom{4}{1}\binom{12}{1} &= \frac{4!}{1!(4-1)!} \cdot \frac{12!}{1!(12-1)!} \\ &= \frac{4 \cdot 3 \cdot 2 \cdot 1}{1 \cdot 3 \cdot 2 \cdot 1} \cdot \frac{12 \cdot 11 \cdot 10 \cdot \ldots \cdot 1}{1 \cdot 11 \cdot 10 \cdot 9 \cdot \ldots \cdot 1} \\ &= 4(12) = 48 \end{aligned}$$

Thus, the probability the dealer draws blackjack is

$$\frac{48}{1326} \approx .036$$

b. For the player to win with blackjack, the player must draw blackjack while the dealer does not draw blackjack.

The probability the player draws blackjack is $\frac{48}{1326}$

Given the player draws blackjack, the number of ways the dealer can draw two cards is:

$$\binom{50}{2} = \frac{50!}{2!(50-2)!} = \frac{50 \cdot 49 \cdot 48 \cdot \ldots \cdot 1}{2 \cdot 1 \cdot 48 \cdot 47 \cdot 46 \cdot \ldots \cdot 1} = 1225$$

Given the player draws blackjack, the number of ways the dealer cannot draw blackjack is:

$$1225 - \binom{3}{1}\binom{11}{1} = 1225 - \frac{3!}{1!(3-1)!} \cdot \frac{11!}{1!(11-1)!}$$

$$= 1225 - \frac{3 \cdot 2 \cdot 1}{1 \cdot 2 \cdot 1} \cdot \frac{11 \cdot 10 \cdot 9 \cdot \ldots \cdot 1}{1 \cdot 10 \cdot 9 \cdot 8 \cdot \ldots \cdot 1}$$

$$= 1225 - 3(11) = 1192$$

Thus, given the player draws blackjack, the probability the dealer will not draw blackjack is $\frac{1192}{1225}$.

The probability the player wins with blackjack is the probability the player draws blackjack and the dealer does not, which is

$$\frac{48}{1326} \cdot \frac{1192}{1225} \approx .035$$

3.91 a. Define the following events:

A_1: {component 1 works properly}
A_2: {component 2 works properly}
B_3: {component 3 works properly}
B_4: {component 4 works properly}
A: {subsystem A works properly}
B: {subsystem B works properly}

The probability a component fails is .1, so the probability a component works properly is $1 - .1 = .9$.

Subsystem A works properly if both components 1 and 2 work properly.

$$P(A) = P(A_1 \cap A_2) = P(A_1)P(A_2) = .9(.9) = .81$$
(since the components operate independently)

Similarly, $P(B) = P(B_1 \cap B_2) = P(B_1)P(B_2) = .9(.9) = .81$

The system operates properly if either subsystem A or B operates properly.

The probability the system operates properly is:

$$P(A \cup B) = P(A) + P(B) - P(A \cap B) = P(A) + P(B) - P(A)P(B)$$
$$= .81 + .81 - .81(.81) = .9639$$

b. The probability exactly 1 subsystem fails is

$$\begin{aligned} P(A \cap B') + P(A' \cap B) &= P(A)P(B') + P(A')P(B) \\ &= .81(1 - .81) + (1 - .81).81 \\ &= .1539 + .1539 = .3078 \end{aligned}$$

c. The probability the system fails is the probability that both subsystems fail or

$$P(A' \cap B') = P(A')P(B') = (1 - .81)(1 - .81) = .0361$$

d. The system operates correctly 99% of the time means it fails 1% of the time. The probability 1 subsystem fails is .19. The probability n subsystems fail is $.19^n$. Thus, we must find n such that

$$.19^n \leq .01$$

Thus, n = 3.

3.93 a. Define the following events:

I_1: {first worker has been giving illegal deductions}
I_2: {second worker has been giving illegal deductions}

$$P(I_1) = \frac{6}{30} = .2$$

b. $P(I_2 | I_1) = \frac{5}{29} \approx .172$

c. We want to find $P(I_1' \cap I_2')$

$$P(I_1') = \frac{24}{30}$$

$$P(I_2' | I_1') = \frac{23}{29}$$

$$P(I_1' \cap I_2') = P(I_2' | I_1')P(I_1') = \frac{24}{30} \cdot \frac{23}{29} \approx .634$$

CHAPTER 4

RANDOM VARIABLES AND PROBABILITY DISTRIBUTIONS

4.1 A random variable is a rule that assigns one and only one value to each simple event of an experiment.

4.3 a. Since we can count the number of words spelled correctly, the random variable is discrete.

b. Since we can assume values in an interval, the amount of liquid waste a plant purifies is continuous.

c. Since time is measured on an interval, the random variable is continuous.

d. Since we can count the number of bacteria per cubic centimeter in drinking water, this random variable is discrete.

e. Since we cannot count the amount of carbon monoxide produced per gallon of unleaded gas, this random variable is continuous.

f. Since weight is measured on an interval, weight is a continuous random variable.

4.5 a. The length of time until recovery is continuous because it lies within an interval.

b. Since we can count the number of violent crimes, this random variable is discrete.

c. Since we can count the number of near misses in a month, this variable is discrete.

d. Since we can count the number of winners each week, this variable is discrete.

e. Since blood pressure lies within some interval, this variable is continuous.

f. Since height lies in some interval, this is also a continuous random variable.

4.7 a. This is *not* a valid distribution because $\sum p(x) \neq 1$.

b. This is a *valid* distribution because $p(x) \geq 0$ for all values of x and $\sum p(x) = 1$.

c. This is *not* a valid distribution because one of the probabilities is negative.

d. The sum of the probabilities over all possible values of the random variable is greater than 1, so this is *not* a valid probability distribution.

4.9 a. $\mu = E(x) = \sum xp(x)$

$$= 10(.1) + 20(.25) + 30(.3) + 40(.2) + 50(.1) + 60(.05)$$
$$= 1 + 5 + 9 + 8 + 5 + 3 = 31$$

$$\sigma^2 = E[(x - \mu)^2] = \sum(x - \mu)^2 p(x)$$
$$= (10 - 31)^2(.1) + (20 - 31)^2(.25) + (30 - 31)^2(.3) + (40 - 31)^2(.2) + (50 - 31)^2(.1) + (60 - 31)^2(.05)$$
$$= 44.1 + 30.25 + .3 + 16.2 + 36.1 + 42.05$$
$$= 169$$

$$\sigma = \sqrt{169} = 13$$

b

c. $\mu \pm 2\sigma \Rightarrow 31 \pm 2(13) \Rightarrow 31 \pm 26 \Rightarrow (5, 57)$

$$P(5 < x < 57) = p(10) + p(20) + p(30) + p(40) + p(50)$$
$$= .10 + .25 + .30 + .20 + .10 = .95$$

4.11 a. The simple events are (where H = head, T = tail)

	HHH	HHT	HTH	THH	HTT	THT	TTH	TTT
x = # heads	3	2	2	2	1	1	1	0

b. If each event is equally likely, then P(simple event) $= \frac{1}{n} = \frac{1}{8}$.

$p(3) = \frac{1}{8}$, $p(2) = \frac{1}{8} + \frac{1}{8} + \frac{1}{8} = \frac{3}{8}$, $p(1) = \frac{1}{8} + \frac{1}{8} + \frac{1}{8} = \frac{3}{8}$, and

$p(0) = \frac{1}{8}$

c.

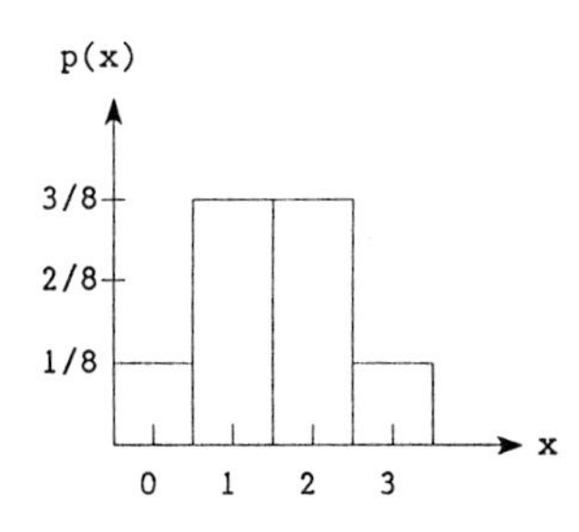

d. $P(x = 2 \text{ or } x = 3) = p(2) + p(3) = \frac{3}{8} + \frac{1}{8} = \frac{4}{8} = \frac{1}{2}$

e. $\mu = E(x) = \sum xp(x) = 0\left(\frac{1}{8}\right) + 1\left(\frac{3}{8}\right) + 2\left(\frac{3}{8}\right) + 3\left(\frac{1}{8}\right) = 1.5$

$$\sigma^2 = E[(x - \mu)^2] = \sum(x - \mu)^2 p(x) = (0 - 1.5)^2\left(\frac{1}{8}\right) + (1 - 1.5)^2\left(\frac{3}{8}\right) + (2 - 1.5)^2\left(\frac{3}{8}\right) + (3 - 1.5)^2\left(\frac{1}{8}\right)$$

$$= .28125 + .09375 + .09375 + .28125$$
$$= .75$$

4.13 a. It would seem that the mean of both would be 1 since they both are symmetric distributions centered at 1.

b. P(x) seems more variable since there appears to be greater probability for the two extreme values of 0 and 2 then there is in the distribution of y.

c. For x: $\mu = 0(.3) + 1(.4) + 2(.3) = 0 + .4 + .6 = 1$
$\sigma^2 = (0 - 1)^2(.3) + (1 - 1)^2(.4) + (2 - 1)^2(.3)$
$= .3 + 0 + .3 = .6$

For y: $\mu = 0(.1) + 1(.8) + 2(.1) = 0 + .8 + .2 = 1$
$\sigma^2 = (0 - 1)^2(.1) + (1 - 1)^2(.8) + (2 - 1)^2(.1)$
$= .1 + 0 + .1 = .2$

4.15 a. $p(6) = .118$

b. $p(5) = .302$

c. $P(x \leq 4) = p(0) + p(1) + p(2) + p(3) + p(4)$
$= .001 + .010 + .060 + .185 + .324 = .580$

4.17 Let ℓ = live and d = die. The simple events for this experiment are:

$\ell\ell\ell\ell$	$d\ell\ell\ell$	$d\ell d\ell$	$d\ell dd$
$\ell\ell\ell d$	$\ell\ell dd$	$dd\ell\ell$	$dd\ell d$
$\ell\ell d\ell$	$\ell d\ell d$	$d\ell\ell d$	$ddd\ell$
$\ell d\ell\ell$	$\ell dd\ell$	ℓddd	$dddd$

$p(\ell) = .8$ and $p(d) = .2$

a. $P(x = 0) = P(dddd) = p(d)p(d)p(d)p(d) = .2(.2)(.2)(.2) = .0016$
(because events are independent)

$P(x = 1) = P(ddd\ell) + P(dd\ell d) + P(d\ell dd) + P(\ell ddd)$
$= 4p(d)p(d)p(d)p(\ell) = 3(.2)(.2)(.2)(.8) = .0256$

$P(x = 2) = P(dd\ell\ell) + P(d\ell d\ell) + P(d\ell\ell d) + P(\ell\ell dd) + P(\ell ddd)$
$+ P(\ell dd\ell)$
$= 6p(d)p(d)p(\ell)p(\ell) = 6(.2)(.2)(.8)(.8) = .1536$

$P(x = 3) = P(\ell\ell\ell d) + P(\ell\ell d\ell) + P(\ell d\ell\ell) + P(d\ell\ell\ell)$
$= 4p(\ell)p(\ell)p(\ell)p(d) = 4(.8)(.8)(.8)(.2) = .4096$

$P(x = 4) = P(\ell\ell\ell\ell) = p(\ell)p(\ell)p(\ell)p(\ell) = .8(.8)(.8)(.8) = .4096$

b.

p(x)
.5 .4 .3 .2 .1
0 1 2 3 4 x

c. $P(x < 2) = p(0) + p(1) = .0016 + .0256 = .0272$

4.19 The sample space of the experiment would be:

S: {BBB, BBG, BGB, GBB, GGB, GBG, BGG, GGG}

where B and G represent a boy and girl respectively. Since the female parent always donates an X chromosome, the gender of any child is determined by the chromosome donated by the father; an X chromosome donated by the father will produce a girl, a Y, a boy. Each child therefore has a .5 probability of being either gender (as might be expected). From this, it follows that each of the above simple events has a probability of 1/8. If we let z represent the number of male offspring, the probability distribution of z is:

z	0	1	2	3
p(z)	1/8	3/8	3/8	1/8

Then,

$$P(\text{At least one boy}) = P(z \geq 1) = 1 - P(z = 0) = 1 - 1/8 = 7/8$$

4.21 a. Let x = damages faced by firm. Then x can take on values 0 (it does not rain) and \$300,000 (it does rain).

$P(x = 0) = .7$ and $P(x = 300{,}000) = .3$

The probability distribution is

x	p(x)
0	.7
\$300,000	.3

b. $E(x) = \$0(.7) + \$300{,}000(.3) = \$90{,}000$

4.23 E(profit contribution) = \$-5000(.3) + \$10,000(.4) + \$30,000(.3)
= \$-1500 + \$4000 + \$9000
= \$11,500

The company will market the new line because its expected profit contribution is greater than \$10,000.

4.25 Let x = bookie's earnings per dollar wagered. Then x can take on values \$1 (you lose) and \$-5 (you win). The only way you win is if you pick 3 winners in 3 games. If the probability of picking 1 winner in 1 game is .5, then $P(www) = p(w)p(w)p(w) = .5(.5)(.5) = .125$ (assuming games are independent).

Thus, the probability distribution for x is

x	p(x)
\$1	.875
\$-5	.125

$E(x) = \sum xp(x) = \$1(.875) - \$5(.125) = \$.875 - \$.725 = \$.25$

4.27 a. x is discrete. It can take on only 7 values.

b. This is a binomial distribution.

c. $p(0) = \binom{6}{0}(.4)^0(.6)^{6-0} = \frac{6!}{0!6!}(.4)^0(.6)^6$

$= \frac{6 \cdot 5 \cdot 4 \cdot 3 \cdot 2 \cdot 1}{1 \cdot 6 \cdot 5 \cdot 4 \cdot 3 \cdot 2 \cdot 1}(1)(.046656)$

$= .046656$

$p(1) = \binom{6}{1}(.4)^1(.6)^{6-1} = \frac{6!}{1!5!}(.4)^1(.6)^5 = .186624$

$p(2) = \binom{6}{2}(.4)^2(.6)^{6-2} = \frac{6!}{2!4!}(.4)^2(.6)^4 = .31104$

$p(3) = \binom{6}{3}(.4)^3(.6)^{6-3} = \frac{6!}{3!3!}(.4)^3(.6)^3 = .27648$

$p(4) = \binom{6}{4}(.4)^4(.6)^{6-4} = \frac{6!}{4!2!}(.4)^4(.6)^2 = .13824$

$p(5) = \binom{6}{5}(.4)^5(.6)^{6-5} = \frac{6!}{5!1!}(.4)^5(.6)^1 = .036864$

$p(6) = \binom{6}{6}(.4)^6(.6)^{6-6} = \frac{6!}{6!0!}(.4)^6(.6)^0 = .004096$

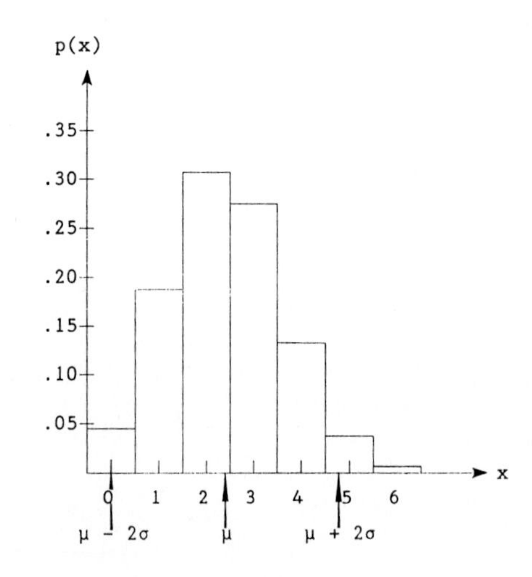

d. $\mu = np = 6(.4) = 2.4$

$\sigma = \sqrt{npq} = \sqrt{6(.4)(.6)} = 1.2$

e. $\mu \pm 2\sigma \Rightarrow 2.4 \pm 2(1.2) \Rightarrow (0, 4.8)$

4.29 a. $p(0) = \binom{4}{0}(.2)^0(.8)^{4-0} = \frac{4!}{0!4!}(.2)^0(.8)^4$

$$= \frac{4 \cdot 3 \cdot 2 \cdot 1}{1 \cdot 4 \cdot 3 \cdot 2 \cdot 1}(1)(.8)^4 = .4096$$

$p(1) = \binom{4}{1}(.2)^1(.8)^{4-1} = \frac{4!}{1!3!}(.2)^1(.8)^3 = .4096$

$p(2) = \binom{4}{2}(.2)^2(.8)^{4-2} = \frac{4!}{2!2!}(.2)^2(.8)^2 = .1536$

$p(3) = \binom{4}{3}(.2)^3(.8)^{4-3} = \frac{4!}{3!1!}(.2)^3(.8)^1 = .0256$

$p(4) = \binom{4}{4}(.2)^4(.8)^{4-4} = \frac{4!}{4!0!}(.2)^4(.8)^0 = .0016$

$p(5) = 0$

b.

x	p(x)
0	.4096
1	.4096
2	.1536
3	.0256
4	.0016

4.31 a. $P(x = 2) = P(x \leq 2) - P(x \leq 1) = .383 - .149 = .234$
(from Table II, Appendix A)

b. $P(x \leq 5) = .151$

c. $P(x > 1) = 1 - P(x \leq 1) = 1 - .737 = .263$

d. $P(x < 10) = P(x \leq 9) = 0$

e. $P(x \geq 10) = 1 - P(x \leq 9) = 1 - .278 = .722$

f. $P(x = 2) = P(x \leq 2) - P(x \leq 1) = .677 - .392 = .285$

4.33 a. $P(x \leq 1) = .005$ (from Table II, Appendix A)

b. $P(x \geq 3) = 1 - P(x \leq 2) = 1 - .027 = .973$

c. $P(x \leq 5) = .403$

d. $P(x < 10) = P(x \leq 9) = .966$

e. $P(x > 10) = 1 - P(x \leq 10) = 1 - .991 = .009$

f. $P(x = 6) = P(x \leq 6) - P(x \leq 5) = .610 - .403 = .207$

4.35 a. No. Since there is a finite number in the population (25,000), the trials are not identical, the trials are not independent, and the probability of success, p, is not constant from trial to trial.

b. Since $n = 20$ is very small compared to the population size ($N = 25,000$), the distribution of x is very close to the binomial distribution. The trials are essentially identical, there are 2 possible outcomes for each trial (schedule rejected or not), the probability of success is essentially constant from trial to trial, and the trials are essentially independent. Thus, x is a binomial random variable.

c. No. The trials would probably not be independent. Also, the chance of a student having his/her schedule rejected in the first 1000 schedules is probably not .2.

d. For (b), $\mu = np = 20(.2) = 4.0$

$\sigma^2 = npq = 20(.2)(.8) = 3.2$

$\sigma = \sqrt{3.2} = 1.789$

4.37 a. Let $n = 15$, $p = .30$, and x = number of cars that will need new gas tanks. The distribution of x is binomial.

$P(x > 10) = 1 - P(x \leq 10) = 1 - .999 = .001$
(from Table II, Appendix A)

b. Let $n = 10,000$, $p = .3$, and x defined as above. This distribution of x is binomial.

$P(x < 3000)$

$\mu = np = 10,000(.3) = 3000$

$\sigma^2 = npq = 10,000(.3)(.7) = 2100$

$\sigma = \sqrt{2100} = 45.826$

The approximate probability is .5.

4.39 Let $n = 20$, $p = .5$, and x = number of correct guesses in 20 trials. Then x has a binomial distribution.

$P(x \geq 9) = 1 - P(x \leq 8) = 1 - .252 = .748$ (from Table II, Appendix A)

4.41 Let $n = 25$, $p = 1 - .9 = .1$ and x = number of invoices that contain errors in 25 trials. Then x has a binomial distribution.

a. $P(x \geq 7) = 1 - P(x \leq 6) = 1 - .991 = .009$
(from Table II, Appendix A)

b. We assume the trials are identical and independent, and the probability of success is constant from trial to trial.

c. Yes. It would be very unlikely (probability = .009) to see 7 or more invoices with errors if $p = .1$. Thus, p is probably greater than .1.

4.43 Let $n = 25$, $p = .1$, and x = number of calculators needing repair in 25 trials. Then x has a binomial distribution.

$P(x \geq 5) = 1 - P(x \leq 4) = 1 - .902 = .098$ (from Table II, Appendix A)

4.45 Let $n = 15$, $p = .5$, and x = number of years of perfect agreement between the January and annual movements in stock prices in 15 trials. Then x has a binomial distribution.

a. $P(x = 15) = P(x \leq 15) - P(x \leq 14) = 1 - 1 = 0$
(from Table II, Appendix A)

b. $P(x \geq 10) = 1 - P(x \leq 9) = 1 - .849 = .151$

4.47 Let $n = 20$, $p = .5$, and x = number of correct questions in 20 trials. Then x has a binomial distribution. We want to find k such that $P(x \geq k) < .05$

or $1 - P(x \leq k - 1) < .05 \Rightarrow P(x \leq k - 1) > .95$
$\Rightarrow k - 1 = 14 \Rightarrow k = 15$
(from Table II, Appendix A)

Note: $P(x \geq 14) = 1 - P(x \leq 13) = 1 - .942 = .058$
$P(x \geq 15) = 1 - P(x \leq 14) = 1 - .979 = .021$

Thus, to have the probability less than .05, the lowest passing grade should be 15.

4.49 a. Let $n = 25$.

i) For $p = 1$, $P(x \leq 2) = 0$
ii) For $p = .8$, $P(x \leq 2) = .000$ (from Table II, Appendix A)
iii) For $p = .5$, $P(x \leq 2) = .000$
iv) For $p = .2$, $P(x \leq 2) = .098$
v) For $p = .05$, $P(x \leq 2) = .873$
vi) For $p = 0$, $P(x \leq 2) = 1$

b.

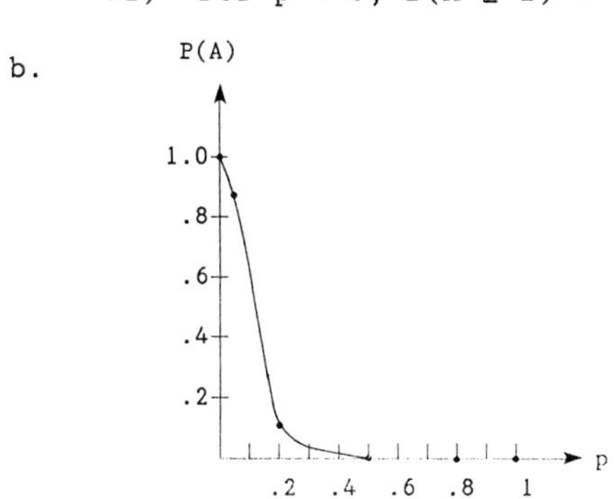

c. i) For $p = 1$, $P(x \leq 3) = 0$
ii) For $p = .8$, $P(x \leq 3) = .000$
iii) For $p = .5$, $P(x \leq 3) = .000$
iv) For $p = .2$, $P(x \leq 3) = .234$
v) For $p = .05$, $P(x \leq 3) = .966$
vi) For $p = 0$, $P(x \leq 3) = 1$

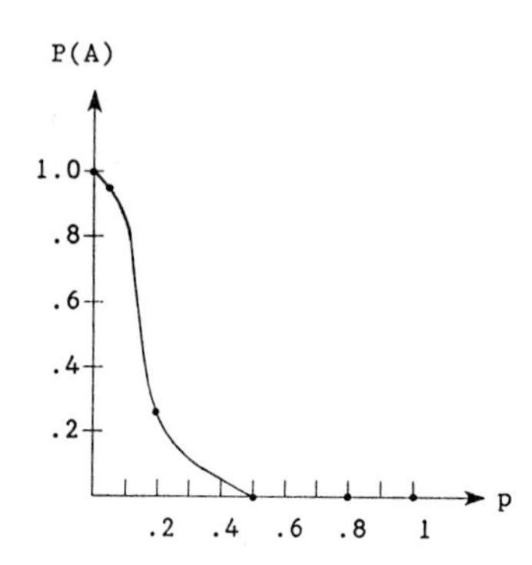

The first plan will detect the defectives more often than the second plan.

4.51 The standard normal random variable z gives the number of standard deviations any normal random variable is from its mean and also the direction (right or left). Any normal random variable can be converted to a standard normal by the formula

$$z = \frac{x - \mu}{\sigma}$$

4.53 Using Table III, Appendix A:

a.

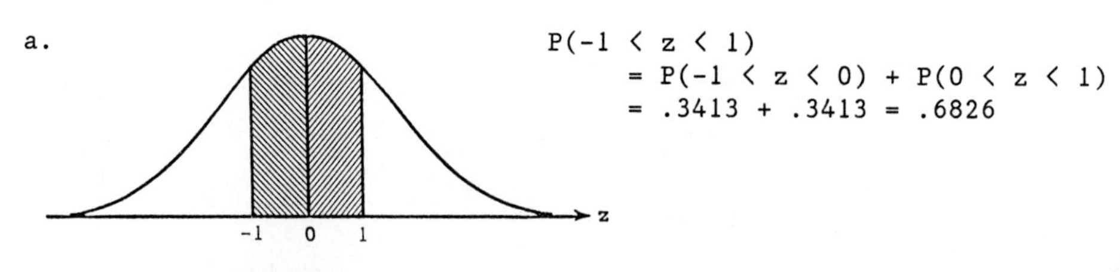

$P(-1 < z < 1)$
$= P(-1 < z < 0) + P(0 < z < 1)$
$= .3413 + .3413 = .6826$

b.

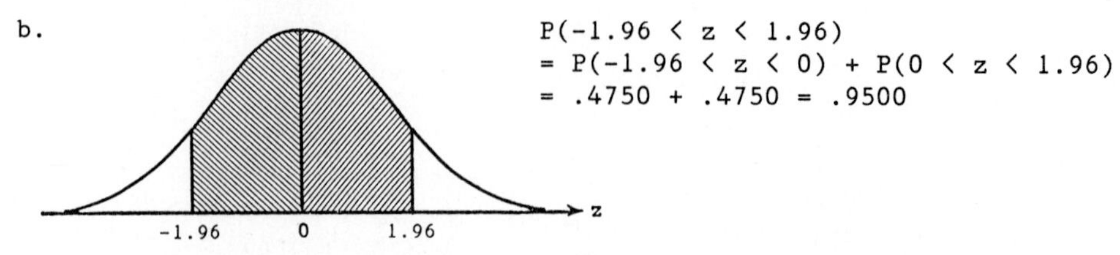

$P(-1.96 < z < 1.96)$
$= P(-1.96 < z < 0) + P(0 < z < 1.96)$
$= .4750 + .4750 = .9500$

c.

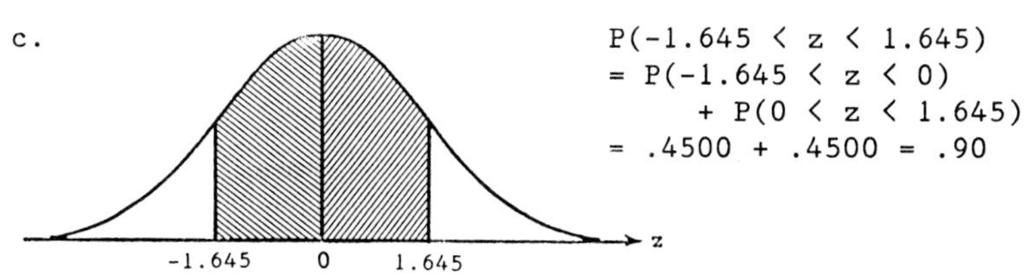

$P(-1.645 < z < 1.645)$
$= P(-1.645 < z < 0)$
$\quad + P(0 < z < 1.645)$
$= .4500 + .4500 = .90$

d.

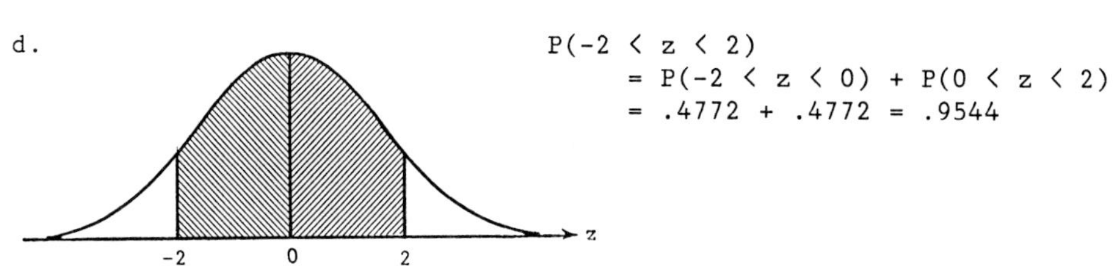

$P(-2 < z < 2)$
$\quad = P(-2 < z < 0) + P(0 < z < 2)$
$\quad = .4772 + .4772 = .9544$

4.55 Using Table III, Appendix A:

a.

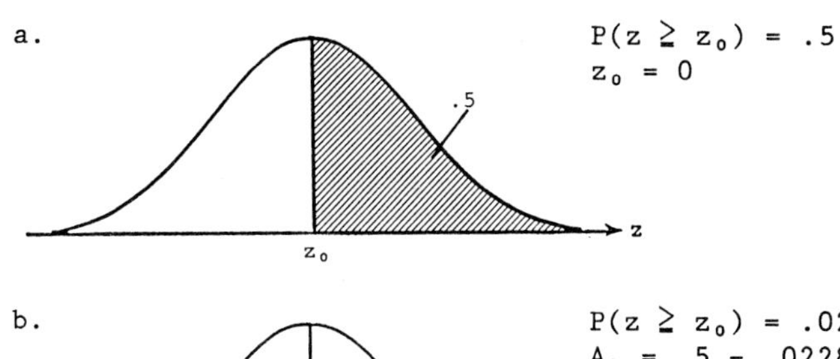

$P(z \geq z_0) = .5$
$z_0 = 0$

b.

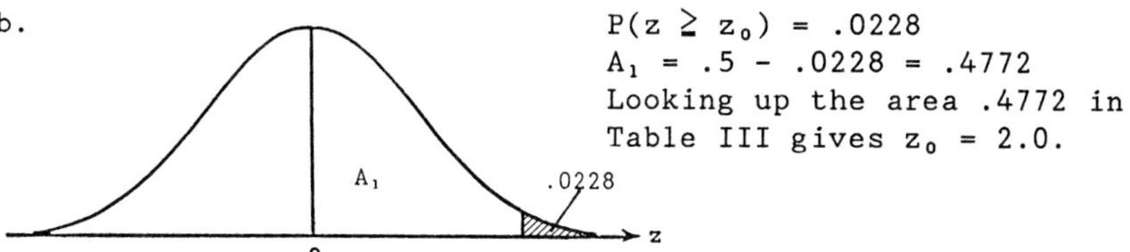

$P(z \geq z_0) = .0228$
$A_1 = .5 - .0228 = .4772$
Looking up the area .4772 in Table III gives $z_0 = 2.0$.

c.

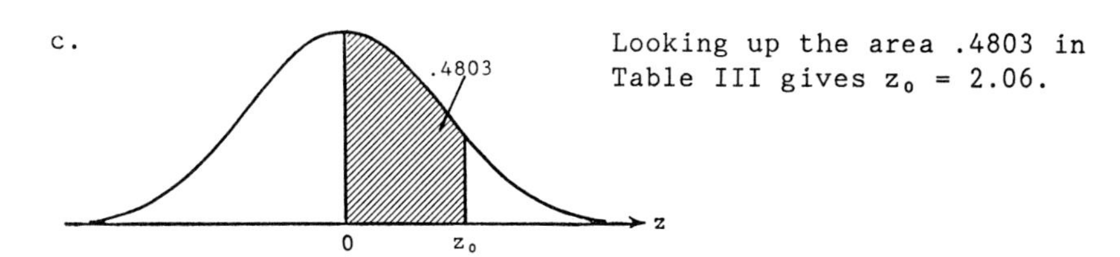

Looking up the area .4803 in Table III gives $z_0 = 2.06$.

d.

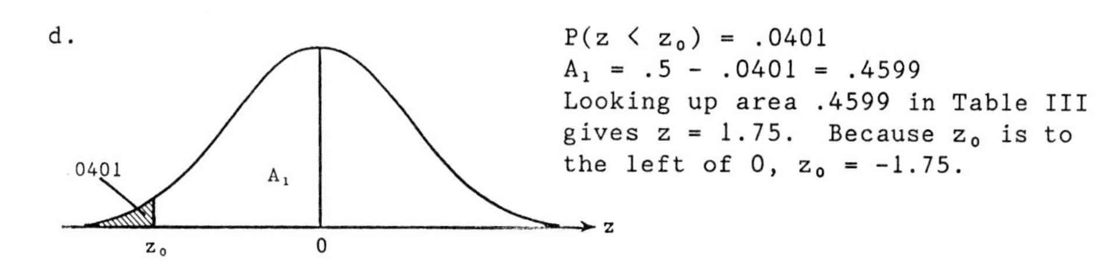

$P(z < z_0) = .0401$
$A_1 = .5 - .0401 = .4599$
Looking up area .4599 in Table III gives $z = 1.75$. Because z_0 is to the left of 0, $z_0 = -1.75$.

4.57 Using the formula $z = \frac{x - \mu}{\sigma}$ with $\mu = 25$ and $\sigma = 5$:

a. $z = \frac{25 - 25}{5} = 0$

b. $z = \frac{30 - 25}{5} = 1$

c. $z = \frac{37.5 - 25}{5} = 2.5$

d. $z = \frac{10 - 25}{5} = -3$

e. $z = \frac{50 - 25}{5} = 5$

f. $z = \frac{32 - 25}{5} = 1.4$

4.59 Using Table III, Appendix A:

a. $P(x \leq 50) = P(z \leq \frac{50 - 45}{10}) = P(z \leq .5) = .5 + .1915 = .6915$

b. $P(x \leq 35.6) = P(z \leq \frac{35.6 - 45}{10}) = P(z \leq .94) = .5 + .3264 = .1736$

c. $P(40.7 \leq x \leq 65.8) = P(\frac{40.7 - 45}{10} \leq z \leq \frac{65.8 - 45}{10})$

$= P(-.43 \leq z \leq 2.08)$

$= .1664 + .4812 = .6476$

d. $P(22.9 \leq x \leq 33.2) = P(\frac{22.9 - 45}{10} \leq z \leq \frac{33.2 - 45}{10})$

$= P(-2.21 \leq z \leq -1.18)$

$= .4864 - .3810 = .1054$

e. $P(x \geq 25.3) = P(z \geq \frac{25.3 - 45}{10}) = P(z \geq -1.97) = .5 + .4756 = .9756$

f. $P(x \leq 25.3) = P(z \leq -1.97) = .5 - .4756 = .0244$

4.61 a. Using Table III, Appendix A, and $\mu = 75$ and $\sigma = 7.5$,

$P(x > 80) = P(z > \frac{80 - 75}{7.5}) = P(z > .67) = .5 - .2486 = .2514$

b. $P(x \leq x_0) = .98$. Find x_0.

$$P(x \leq x_0) = P(z \leq \frac{x_0 - 75}{7.5}) = P(z \leq z_0) = .98$$

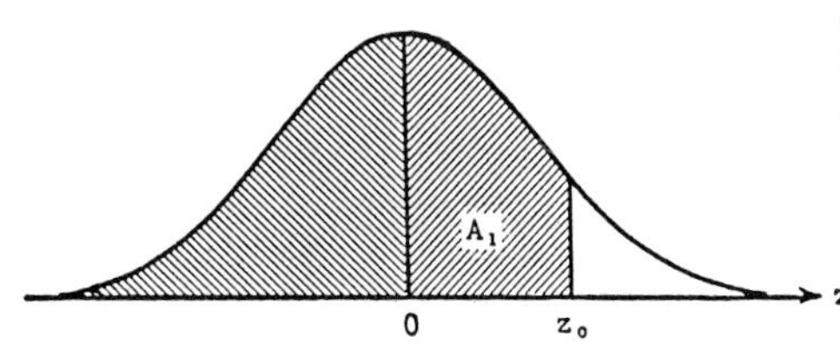

$A_1 = .98 - .5 = .4800$
Looking up area .4800 in Table III, $z_0 = 2.05$.

$$z_0 = \frac{x_0 - 75}{7.5} \Rightarrow 2.05 = \frac{x_0 - 75}{7.5} \Rightarrow x_0 = 90.375$$

4.63 Using Table III, Appendix A, and $\mu = \$600,000$ and $\sigma = \$200,000$:

a. $P(x > 1,000,000) = P(z > \frac{1,000,000 - 600,000}{200,000}) = P(z > 2.00)$

$$= .5 - .4772 = .0228$$

Thus, about 2.28% of the baseball players make more than \$1,000,000 per year.

b. There are a few superstars that have very large salaries, whereas most of the salaries are much smaller. The distribution is probably skewed to the right.

4.65 From Table III, Appendix A, with $\mu = 6.3$ and $\sigma = .6$

a. $P(x < 5) = P(z < \frac{5 - 6.3}{.6}) = P(z < -2.17) = .5 - .4850 = .0150$

Thus, the percentage of days when the oxygen content is undesirable is 1.5%.

b. We would expect the oxygen contents to fall within 2 standard deviations of the mean, or

$$\mu \pm 2\sigma \Rightarrow 6.3 \pm 2(.6) \Rightarrow 6.3 \pm 1.2 \Rightarrow (5.1, 7.5)$$

4.67 From Table III, Appendix A, with $\mu = \$10.50$ and $\sigma = \$1.25$

a. $P(x > 12.00) = P(z > \frac{12 - 10.50}{1.25}) = P(z > 1.2) = .5 - .3849 = .1151$

b. $P(x > 10.00) = P(z > \frac{10 - 10.50}{1.25}) = P(z > -.4) = .5 + .1554 = .6554$

c. $P(x \geq x_m) = P(z \geq \frac{x_m - 10.50}{1.25}) = P(z > z_m) = .5$

Thus, $z_m = 0 \Rightarrow z_m = \frac{x_m - 10.50}{1.25}$

$\Rightarrow 0 = x_m - 10.50 \Rightarrow x_m = 10.50$

The median and mean are both \$10.50.

4.69 From Table III, Appendix A, and $\sigma = .4$:

$P(x > 6) = .01$

$P(x > 6) = P(z > \frac{6 - \mu}{.4}) = P(z > z_0) = .01$

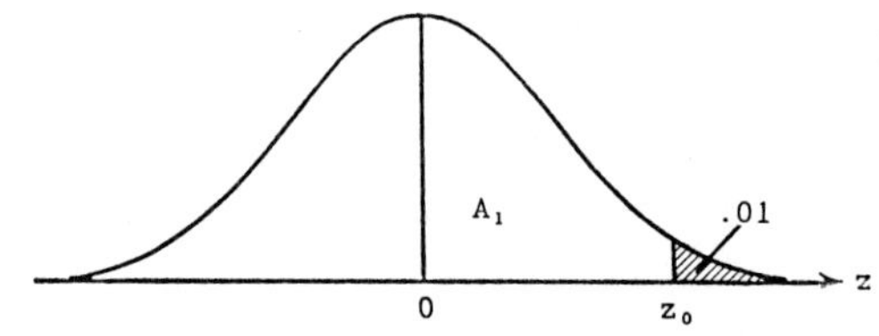

$A_1 = .5 - .01 = .4900$
From Table III, $z_0 = 2.33$

$2.33 = \frac{6 - \mu}{.4} \Rightarrow \mu = 6 - 2.33(.4) = 5.068$

4.71 Using Table III, Appendix A, and $\mu = 435$ and $\sigma = 72$:

$P(x > x_0) = .30$. Find x_0.

$P(x > x_0) = P(z > \frac{x_0 - 435}{72}) = P(z > z_0) = .30$

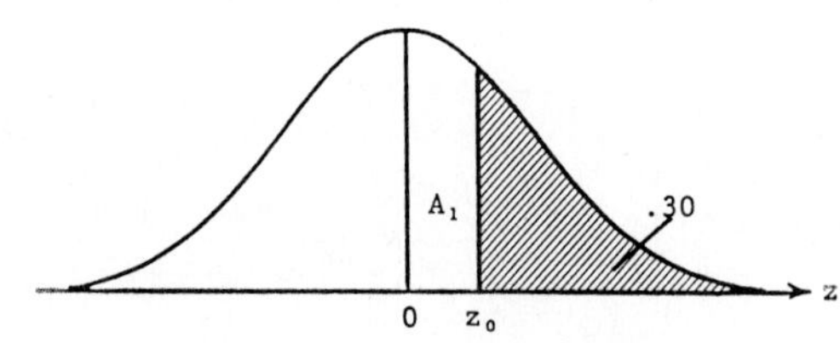

$A_1 = .5 - .3 = .2000$
Using Table III and area .2000,
$z_0 = .52$

$z_0 = \frac{x_0 - 435}{72} \Rightarrow .52 = \frac{x_0 - 435}{72} \Rightarrow x_0 = 472.44 \approx 473$

4.73 Using Table III, Appendix A, and $\mu = 7200$ and $\sigma = 300$:

a. $P(x \leq x_0) = .94$. Find x_0.

$$P(x \leq x_0) = P(z \leq \frac{x_0 - 7200}{300}) = P(z \leq z_0) = .94$$

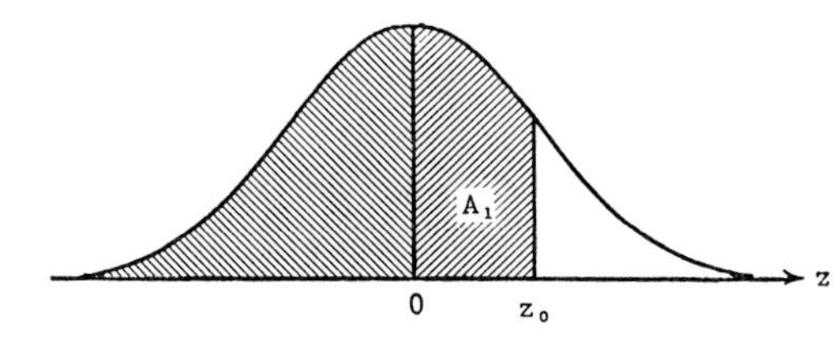

$A_1 = .94 - .50 = .4400$
Using Table III and area .4400,
$z_0 = 1.555$.

$$z_0 = \frac{x_0 - 7200}{300} \Rightarrow 1.555 = \frac{x_0 - 7200}{300} \Rightarrow x_0 = 7666.5 \approx 7667$$

b. If the company produces 7,667 loaves, the company will be left with more than 500 loaves if the demand is less than 7,667 - 500 = 7167.

$$P(x < 7167) = P(z < \frac{7167 - 7200}{300}) - P(z < -.11)$$

$$= .5 - .0438 = .4562 \quad \text{(from Table III, Appendix A)}$$

Thus, on 45.62% of the days the company will be left with more than 500 loaves.

4.75 When n is very large, it is difficult to use the binomial formula and no tables exist. The normal distribution approximates the binomial distribution very well for large n.

4.77 a. $\mu = np = 50(.01) = .5$, $\sigma = \sqrt{npq} = \sqrt{50(.01)(.99)} = .704$

$\mu \pm 3\sigma \Rightarrow .5 \pm 3(.704) \Rightarrow .5 \pm 2.112 \Rightarrow (-1.612, 2.612)$

Since this interval does not fall in the interval (0, n = 50), the normal approximation is not appropriate.

b. $\mu = np = 20(.45) = 9$, $\sigma = \sqrt{npq} = \sqrt{20(.45)(.55)} = 2.225$

$\mu \pm 3\sigma \Rightarrow 9 \pm 3(2.225) \Rightarrow 9 \pm 6.675 \Rightarrow (2.325, 15.675)$

Since this interval falls in the interval (0, n = 20), the normal approximation is appropriate.

c. $\mu = np = 10(.4) = 4$, $\sigma = \sqrt{npq} = \sqrt{10(.4)(.6)} = 2.001$

$\mu \pm 3\sigma \Rightarrow 4 \pm 3(2.001) \Rightarrow 4 \pm 6.003 \Rightarrow (-2.003, 10.003)$

Since this interval does not fall within the interval (0, n = 10), the normal approximation is not appropriate.

d. $\mu = np = 1000(.1) = 100$, $\sigma = \sqrt{npq} = \sqrt{1000(.1)(.9)} = 9.487$

$\mu \pm 3\sigma \Rightarrow 100 \pm 3(9.487) \Rightarrow 100 \pm 28.461 \Rightarrow (71.539, 128.461)$

Since this interval falls within the interval (0, n = 1000), the normal approximation is appropriate.

e. $\mu = np = 200(.8) = 160$, $\sigma = \sqrt{npq} = \sqrt{200(.8)(.2)} = 5.657$

$\mu \pm 3\sigma \Rightarrow 160 \pm 3(5.657) \Rightarrow 160 \pm 16.971 \Rightarrow (143.029, 176.971)$

Since this interval falls within the interval (0, n = 200), the normal approximation is appropriate.

f. $\mu = np = 35(.7) = 24.5$, $\sigma = \sqrt{npq} = \sqrt{35(.7)(.3)} = 2.711$

$\mu \pm 3\sigma \Rightarrow 24.5 \pm 3(2.711) \Rightarrow 24.5 \pm 8.133 \Rightarrow (16.367, 32.633)$

Since this interval falls within the interval (0, n = 35), the normal approximation is appropriate.

4.79 a. Using Table II, $P(x \leq 12) = .500$

$\mu = np = 25(.5) = 12.5$, $\sigma = \sqrt{npq} = \sqrt{25(.5)(.5)} = 2.5$

Using the normal approximation,

$$P(x \leq 12) \approx P(z \leq \frac{12 + .5 - 12.5}{2.5}) = P(z \leq 0) = .5$$

b. Using Table II, $P(x \geq 15) = 1 - P(x \leq 14) = 1 - .788 = .212$

Using the normal approximation,

$$P(x \geq 15) \approx P(z \geq \frac{15 - .5 - 12.5}{2.5}) = P(z \geq .8) = .5 - .2881$$

$$= .2119$$

(from Table III, Appendix A)

c. Using Table II, $P(9 \leq x \leq 15) = P(x \leq 15) - P(x \leq 8)$

$= .885 - .054 = .831$

Using the normal approximation,

$$P(9 \leq x \leq 15) \approx P\left(\frac{9 - .5 - 12.5}{2.5} \leq z \leq \frac{15 + .5 - 12.5}{2.5}\right)$$

$$= P(-1.6 \leq z \leq 1.2)$$

$$= .4452 + .3849 = .8301$$

(from Table III, Appendix A)

4.81 $\mu = np = 1000(.5) = 500$, $\sigma = \sqrt{npq} = \sqrt{1000(.5)(.5)} = 15.811$

a. Using the normal approximation,

$$P(x > 500) \approx P\left(z > \frac{500 + .5 - 500}{15.811}\right) = P(z > .03) = .5 - .0120$$

$$= .4880$$

(from Table III, Appendix A)

b. $$P(490 \leq x < 500) \approx P\left(\frac{490 - .5 - 500}{15.811} \leq z < \frac{500 - .5 - 500}{15.811}\right)$$

$$= P(-.66 \leq z < -.03)$$

$$= .2454 - .0120 = .2334$$

(from Table III, Appendix A)

c. $$P(x > 1000) \approx P\left(z > \frac{1000 + .5 - 500}{15.811}\right) = P(z > 31.66) = .5 - .5 = 0$$

(from Table III, Appendix A)

4.83 a. For an absentee rate of 20%, $p = .2$ and $n = 50$.

$\mu = np = 50(.2) = 10$ and $\sigma = \sqrt{npq} = \sqrt{50(.2)(.8)} = 2.828$

90% of 50 is 45. If at least 45 are on the job, then at most 5 are absent.

$$P(x \leq 5) \approx P\left(z \leq \frac{5 + .5 - 10}{2.828}\right) = P(z \leq -1.59) = .5 - .4441 = .0559$$

(from Table III, Appendix A)

b. $\mu \pm 3\sigma \Rightarrow 10 \pm 3(2.828) \Rightarrow 10 \pm 8.484 \Rightarrow (1.516, 18.484)$

Since this interval falls within $(0, n = 50)$, the normal approximation is appropriate.

c. For $p = .02$, $\mu = np + 50(.02) = 1$ and $\sigma = \sqrt{npq} = \sqrt{50(.02)(.98)}$ $= .99$

$\mu \pm 3\sigma \Rightarrow 1 \pm 3(.99) \Rightarrow 1 \pm 2.97 \Rightarrow (-1.97, 3.97)$

Since this interval does not fall within the interval $(0, n = 50)$, the normal approximation is not appropriate.

4.85 a. $n = 328$, $p = .10$. $\mu = np = 328(.1) = 32.8$, and

$\sigma = \sqrt{npq} = \sqrt{328(.1)(.9)} = 5.433$

$$P(x \geq 50) \approx P\left(z \geq \frac{50 - .5 - 32.8}{5.433}\right) = P(z \geq 3.07)$$
$$= .5 - .4989 = .0011$$
(from Table III, Appendix A)

b. $$P(x \geq 184) \approx P\left(z \geq \frac{184 - .5 - 32.8}{5.433}\right) = P(z \geq 27.74) = .5 - .5 = 0$$

The probability of observing this event or anything rarer is essentially 0. This is extremely rare.

c. Since the probability of observing this event is essentially 0 if $p = .1$, we would conclude that p is actually greater than .1 or more than 10% of the leases are salable.

4.87 a. $n = 15{,}000$ and $p = .45$

$E(x) = \mu = np = 15{,}000(.45) = 6750$

$\sigma^2 = npq = 15{,}000(.45)(.55) = 3712.5$

b. $\sigma = \sqrt{3712.5} = 60.93$

Using the normal approximation to the binomial,

$$P(x > 6900) \approx P\left(z > \frac{6900 + .5 - 6750}{60.93}\right) = P(z > 2.47) = .5 - .4932$$
$$= .0068$$
(from Table III, Appendix A)

c. Using the normal approximation to the binomial,

$$P(x > 7000) = P\left(z > \frac{7000 + .5 - 6750}{60.93}\right) = P(z > 4.11) \approx .5 - .5 = 0$$

Since this probability is essentially 0, we would not expect to see more than 7000 deaths in any one year.

4.89 a. Let x = number of households that own PC's in the 500 sampled.

$\mu = np = 500(.60) = 300$, $\sigma = \sqrt{npq} = \sqrt{500(.60)(.40)} = 10.954$

Using the normal approximation to the binomial,

$$P(x \geq 325) \approx P(z \geq \frac{325 - .5 - 300}{10.954}) = P(z \geq 2.24) = .5 - .4875 = .0125$$

(from Table III, Appendix A)

b. Yes. The probability of seeing 325 or more households with PC's is .0125 if the 60% figure is correct. This probability is quite small. Thus, we would conclude that more than 60% of this community's households own PC's.

4.91 Let x = number of incoming freshmen who graduate in 4 years.

$\mu = np = 200(.57) = 114$, $\sigma = \sqrt{npq} = \sqrt{200(.57)(.43)} = 7.00$

a. Using the normal approximation to the binomial,

$$P(x \geq 100) \approx P(z \geq \frac{100 - .5 - 114}{7.00}) = P(z \geq -2.07) = .5 + .4808 = .9808$$

(Using Table III, Appendix A)

b. $$P(40 < x < 80) \approx P(\frac{40 + .5 - 114}{7} < z < \frac{80 - .5 - 114}{7})$$

$$= P(-10.5 < z < -4.93)$$
$$\approx .5 - .5 = 0$$ (Using Table III, Appendix A)

4.93 Let x = number of patients who wait more than 20 minutes.

a. $\mu = np = 150(.5) = 75$, $\sigma = \sqrt{npq} = \sqrt{150(.5)(.5)} = 6.124$

$$P(x > 75) \approx P(z > \frac{75 + .5 - 75}{6.124}) = P(z > .08) = .5 - .0319 = .4681$$

(from Table III, Appendix A)

b. $$P(x > 85) \approx P(z > \frac{85 + .5 - 75}{6.124}) = P(z > 1.71) = .5 - .4564 = .0436$$

(from Table III, Appendix A)

c. $$P(60 < x < 90) \approx P(\frac{60 + .5 - 75}{6.124} < z < \frac{90 - .5 - 75}{6.124})$$

$$= P(-2.37 < z < 2.37) = .4911 + .4911 = .9822$$

(from Table III, Appendix A)

4.95 a. Number of damaged items is discrete.

b. The average monthly sales is continuous.

c. The number of square feet is continuous.

d. The length of time is continuous.

4.97 a. $\mu = E(x) = \sum xp(x) = 10(.2) + 12(.2) + 18(.1) + 20(.5) = 16.2$

$$\sigma^2 = E(x^2) - \mu^2 = \sum x^2p(x) - \mu^2$$
$$= 10^2(.2) + 12^2(.2) + 18^2(.1) + 20^2(.5) - 16.2^2$$
$$= 281.2 - 262.44 = 18.76$$
$$\sigma = \sqrt{18.76} = 4.331$$

b. $P(x < 15) = P(10) + P(12) = .2 + .2 = .4$

c. $\mu \pm 2\sigma \Rightarrow 16.2 \pm 2(4.331) \Rightarrow 16.2 \pm 8.662 \Rightarrow (7.538, 24.862)$

d. $P(7.538 < x < 24.862) = P(10) + P(12) + P(18) + P(20)$
$= .2 + .2 + .1 + .5 = 1.0$

4.99 Using Table III, Appendix A:

a. $P(z \leq 1.1) = .5 + .3643 = .8643$

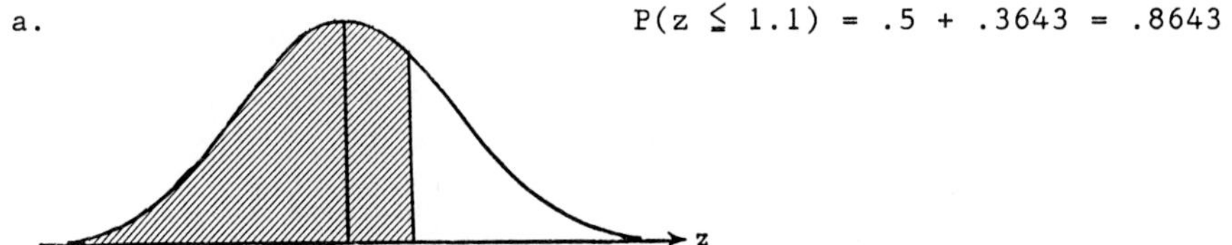

b. $P(z \geq 1.1) = .5 - .3643 = .1357$

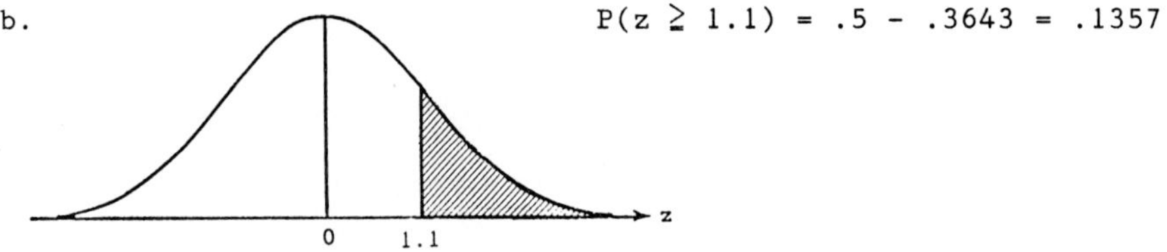

c. $P(z \geq -1.86) = .5 + .4686 = .9686$

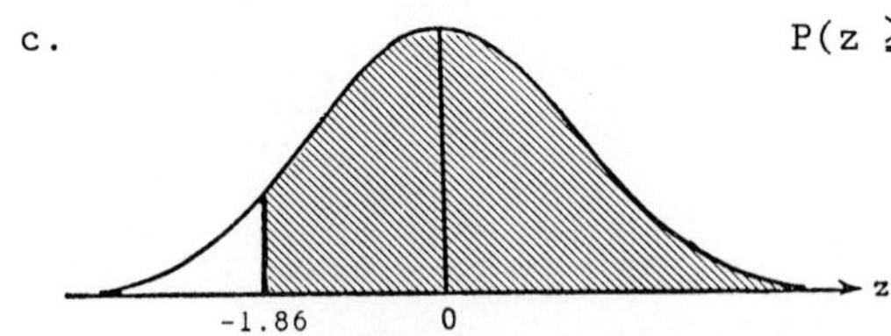

d. $P(-2.75 \leq z \leq -.55)$
$= P(.55 \leq z \leq 2.75)$
$= .4970 - .2088$
$= .2882$

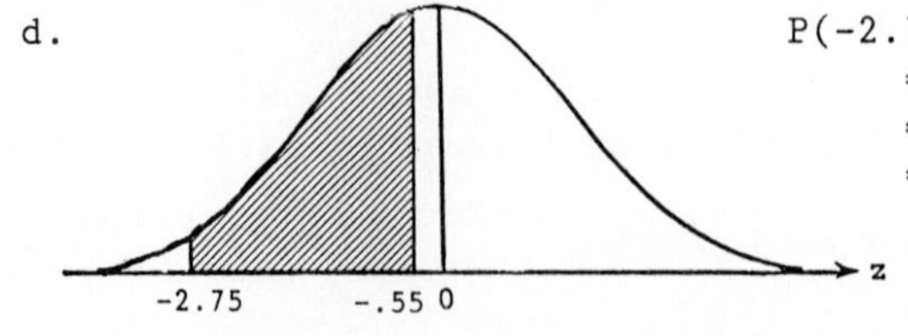

e.

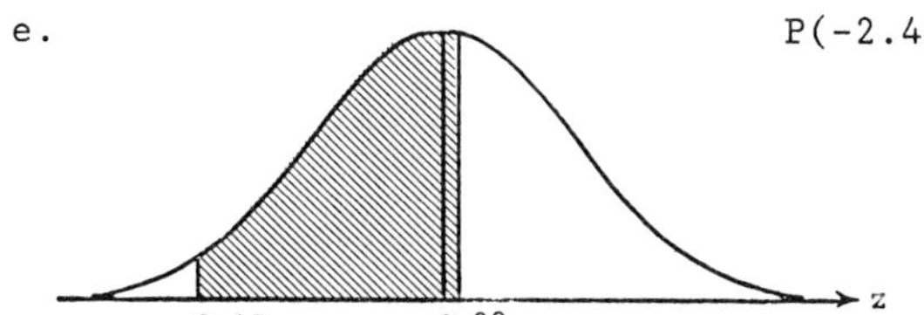

$P(-2.45 \leq z \leq .38) = .4929 + .1480$
$= .6409$

f.

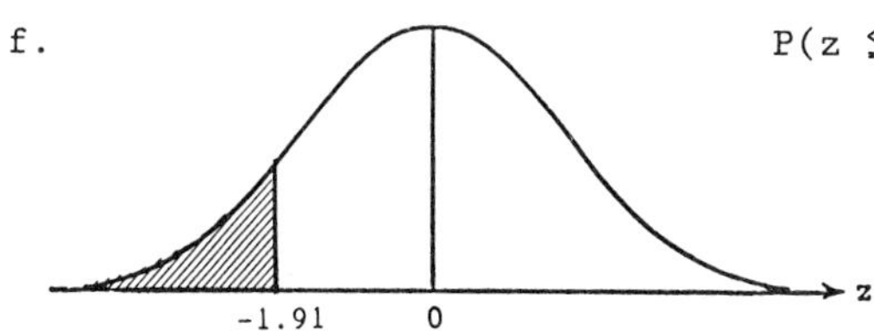

$P(z \leq -1.91) = P(z \geq 1.91)$
$= .5 - .4719 = .0281$

4.101 Using Table III, Appendix A:

a.

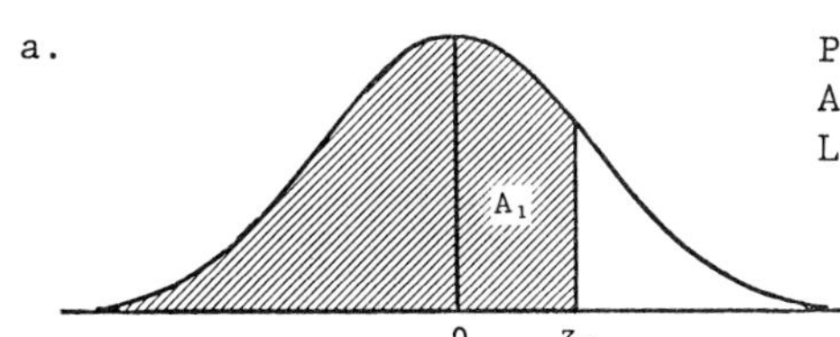

$P(z \leq z_0) = .8023$
$A_1 = .8023 - .5 = .3023$
Looking up area .3023, $z_0 = .85$

b.

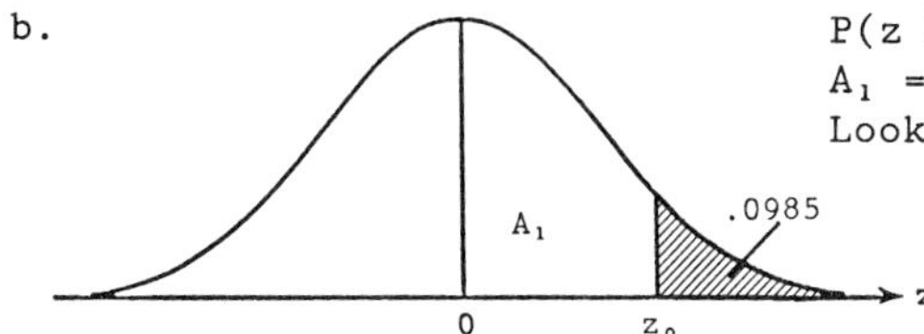

$P(z \geq z_0) = .0985$
$A_1 = .5 - .0985 = .4015$
Looking up area .4015, $z_0 = 1.29$

c.

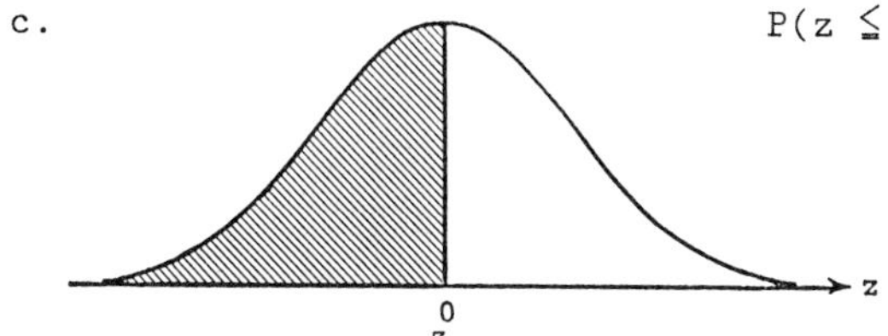

$P(z \leq z_0) = .5 \Rightarrow z_0 = 0$

d.

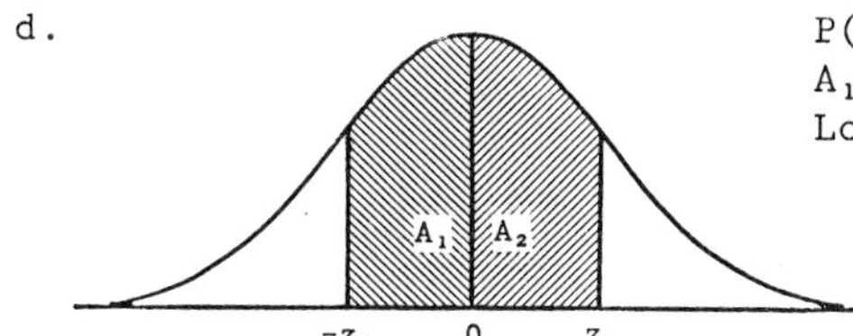

$P(-z_0 \leq z \leq z_0) = .6212$
$A_1 = A_2 = .6212/2 = .3106$
Looking up area .3106, $z_0 = .88$

e.

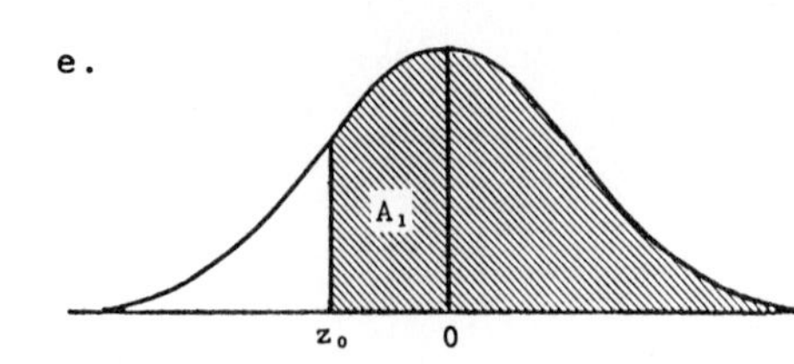

$P(z \geq z_0) = .8508$
$A_1 = .8508 - .5 = .3508$
Looking up area .3508, $z = 1.04$.
Since z_0 is to the left of 0,
$z_0 = -1.04$.

f.

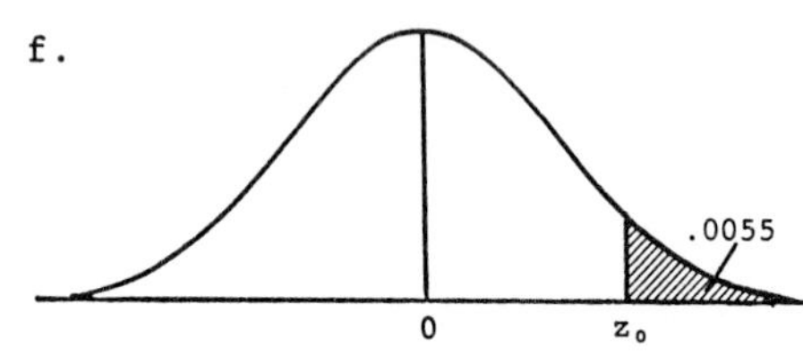

$P(z \geq z_0) = .0055$
$A_1 = .5 - .0055 = .4945$
Looking up area .4945, $z_0 = 2.54$.

4.103 Using Table III, Appendix A:

a. $P(x \geq x_0) = .5$. Find x_0.

$$P(x \geq x_0) = P\left(z \geq \frac{x_0 - 60}{8}\right) = P(z \geq z_0) = .5 \Rightarrow z_0 = 0$$

$$z_0 = \frac{x_0 - 60}{8} \Rightarrow 0 = \frac{x_0 - 60}{8} \Rightarrow x_0 = 60$$

b. $P(x \leq x_0) = .9911$. Find x_0.

$$P(x \leq x_0) = P\left(z \leq \frac{x_0 - 60}{8}\right) = P(z \leq z_0) = .9911$$

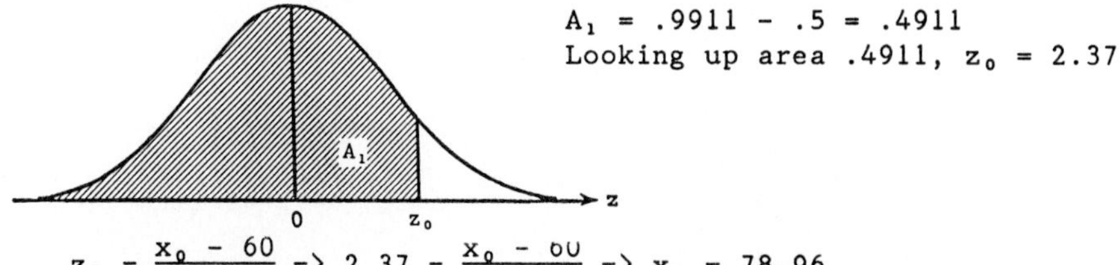

$A_1 = .9911 - .5 = .4911$
Looking up area .4911, $z_0 = 2.37$

$$z_0 = \frac{x_0 - 60}{8} \Rightarrow 2.37 = \frac{x_0 - 60}{8} \Rightarrow x_0 = 78.96$$

c. $P(x \leq x_0) = .0028$. Find x_0.

$$P(x \leq x_0) = P(z \leq \frac{x_0 - 60}{8}) = P(z \leq z_0) = .0028$$

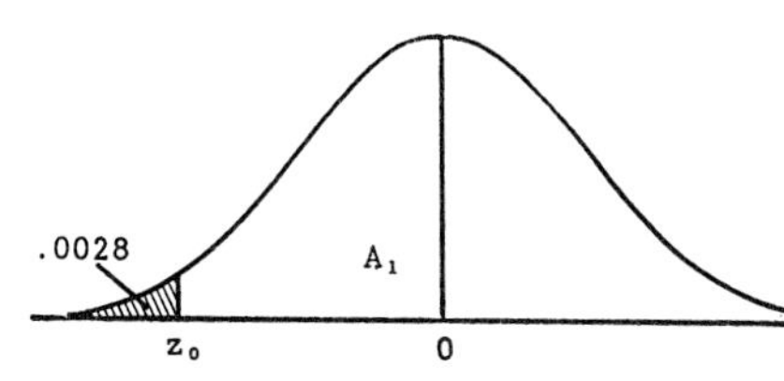

$A_1 = .5 - .0028 = .4972$
Looking up area .4972, $z_0 = 2.77$
Since z_0 is to the left of 0,
$z_0 = -2.77$.

$$z_0 = \frac{x_0 - 60}{8} \Rightarrow -2.77 = \frac{x_0 - 60}{8} \Rightarrow x_0 = 37.84$$

d. $P(x \geq x_0) = .0228$. Find x_0.

$$P(x \geq x_0) = P(z \geq \frac{x_0 - 60}{8}) = P(z \geq z_0) = .0228$$

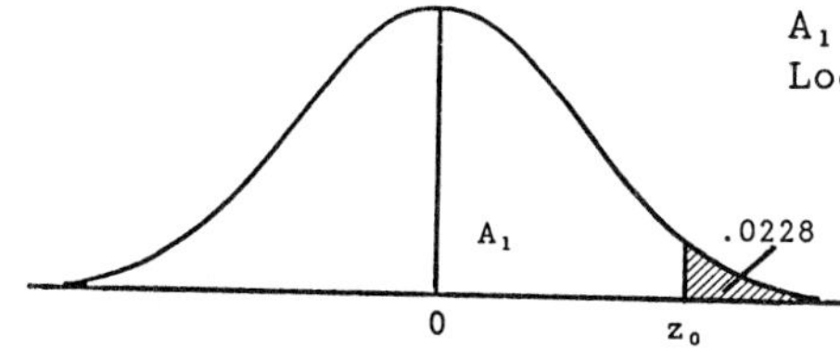

$A_1 = .5 - .0228 = .4772$
Looking up area .4772, $z_0 = 2.0$

$$z_0 = \frac{x_0 - 60}{8} \Rightarrow 2 = \frac{x_0 - 60}{8} \Rightarrow x_0 = 76$$

e. $P(x \leq x_0) = .1003$. Find x_0.

$$P(x \leq x_0) = P(z \leq \frac{x_0 - 60}{8}) = P(z \leq z_0) = .1003$$

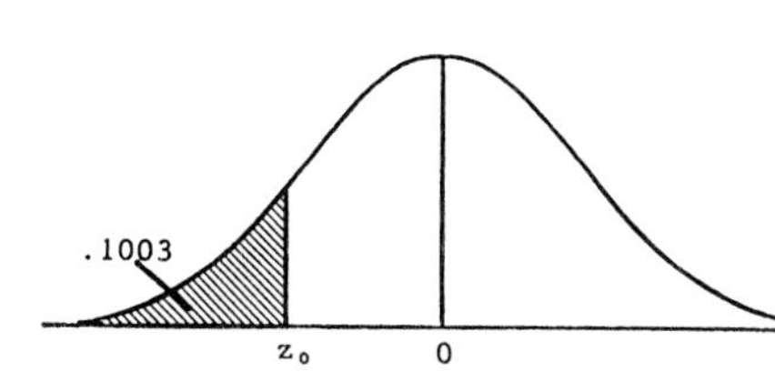

$A_1 = .5 - .1003 = .3997$
Looking up area .3997, $z = 1.28$.
Since z_0 is to the left of 0,
$z_0 = -1.28$.

$$z_0 = \frac{x_0 - 60}{8} \Rightarrow -1.28 = \frac{x_0 - 60}{8} \Rightarrow x_0 = 49.76$$

f. $P(x \geq x_0) = .7995$. Find x_0.

$$P(x \geq x_0) = P(z \geq \frac{x_0 - 60}{8}) = P(z \geq z_0) = .7995$$

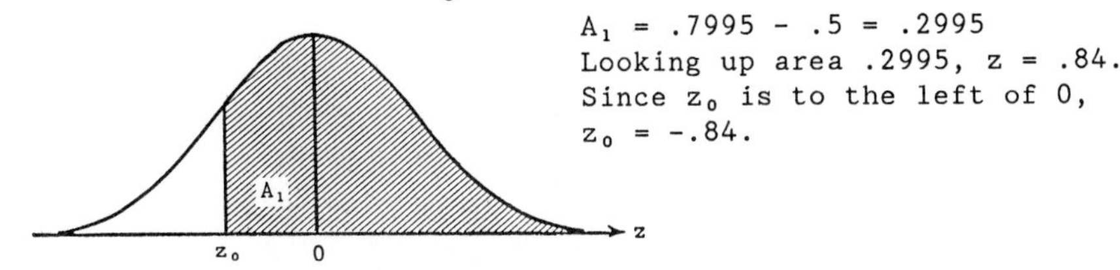

$A_1 = .7995 - .5 = .2995$
Looking up area .2995, z = .84.
Since z_0 is to the left of 0,
$z_0 = -.84$.

$$z_0 = \frac{x_0 - 60}{8} \Rightarrow -.84 = \frac{x_0 - 60}{8} \Rightarrow x_0 = 53.28$$

4.105 Let x = score on mathematics achievement test

$$P(x \geq 70) = P(z \geq \frac{70 - 77}{7.3}) = P(z \geq -.96) = .5 + .3315 = .8315$$

(Using Table III, Appendix A)

Thus, .8315 x 100% = 83.15% will pass the first time.

4.107
$$E(x) = \sum xp(x) = 10{,}000(.05) + 15{,}000(.15) + 20{,}000(.35) + 25{,}000(.25) + 30{,}000(.15) + 35{,}000(.05) = 22{,}250$$

4.109 Let x = tensile strength of a metal part.

Using Table III, Appendix A:

$$P(x > 30) = P(z > \frac{30 - 25}{2}) = P(z > 2.5) = .5 - .4938 = .0062$$

$$P(21 \leq x \leq 30) = P(\frac{21 - 25}{2} \leq z \leq 2.5) = P(-2 \leq z \leq 2.5) = .4772 + .4938 = .9710$$

$$P(x < 21) = P(z < -2) = .5 - .4772 = .0228.$$

Let y = profit per metal part. The probability distribution for y is

y	p(y)
-2	.0228
-1	.0062
10	.9710
	1.0000

$$E(y) = -2(.0228) - 1(.0062) + 10(.9710) = 9.6582 \approx \$9.66 \text{ per part}$$

4.111 a.
$$P(x \leq 1000) = P(x = 0) + P(x = 500) + P(x = 1000) = .01 + .22 + .30 = .53$$

b. $E(x) = \sum xp(x) = 0(.01) + 500(.22) + 1000(.30) + 1500(.22) + 2000(.25)$
$= \$1,240$

c. Profit = Pay back price - cost

E(Profit) = E(x - cost) = E(x) - 1000 = \$1,240 - 1000 = \$240

d. $\sigma^2 = E[(x - \mu)^2] = \sum(x - \mu)^2 p(x)$

$= (0 - 1240)^2(.01) + (500 - 1240)^2(.22) + (1000 - 1240)^2(.30) + (1500 - 1240)^2(.22) + (2000 - 1240)^2(.25)$

$= 15,376 + 120,472 + 17,280 + 14,872 + 144,440 = \$312,400$

$\sigma = \sqrt{312,400} = 558.9275$

4.113 Let x = number of people responding to a questionnaire in 20 trials. Then x is a binomial random variable with n = 20 and p = .4.

a. $P(x > 12) = 1 - p(x \leq 12) = 1 - .979 = .021$
(Table II, Appendix A)

b. We know from the Empirical Rule that almost all the observations are larger than $\mu - 2\sigma$. ($\approx$ 95% are between $\mu - 2\sigma$ and $\mu + 2\sigma$). Thus $\mu - 2\sigma > 100$.

For the binomial, $\mu = np = n(.4)$ and $\sigma = \sqrt{npq} = \sqrt{n(.4)(.6)}$

$\mu - 2\sigma > 100 \Rightarrow .4n - 2\sqrt{.24n} > 100 \Rightarrow .4n - .98\sqrt{n} - 100 > 0$

Solving for $\sqrt{n}$, we get

$$\sqrt{n} = \frac{.98 \pm \sqrt{.98^2 - 4(.4)(-100)}}{2(.4)} = \frac{.98 \pm 12.687}{.8}$$

$$\Rightarrow \sqrt{n} = 17.084 \Rightarrow n = 17.084^2 = 291.9 \approx 292$$

4.115 Let x = time to complete race. Using Table III, Appendix A:

a. $P(34 < x < 60) = P(\frac{34 - 49}{8} < z < \frac{60 - 49}{8}) = P(-1.88 < z < 1.38)$
$= .4699 + .4162$
$= .8861$

(If you interpolate, this becomes $P(-1.875 < z < 1.375)$ $= .4696 + .4154 = .8850$)

b. $P(x < 34) = P(z < -1.88) = .5 - .4699 = .0301$

4.117 a. Let x = number of hoaxes in 5 calls.

$n = 5,\ p = \frac{1}{6}$

$$P(x = 0) = \binom{5}{0}\left(\frac{1}{6}\right)^{0}\left(\frac{5}{6}\right)^{5-0} = \frac{5!}{0!5!}\left(\frac{5}{6}\right)^{5} = .4019$$

b. If three need assistance, then 5 - 3 = 2 were hoaxes.

$$P(x = 2) = \binom{5}{2}\left(\frac{1}{6}\right)^{2}\left(\frac{5}{6}\right)^{5-2} = \frac{5!}{2!3!}\left(\frac{1}{6}\right)^{2}\left(\frac{5}{6}\right)^{3} = .1608$$

c. We must assume

1. The trials are identical
2. The probability of a hoax = 1/6 is constant from trial to trial.
3. The trials are independent.

d. The expected number of hoaxes = $E(x) = np = 10000(1/6) = 1666.67$. At \$30 per call, the amount of money wasted is

$$30E(x) = 30(1666.67) = \$50,000$$

4.119 Let x = amount of tar. Using Table III, Appendix A:

a. $P(x > 10) = P(z > \frac{10 - 8}{1.9}) = P(z > 1.05) = .5 - .3531 = .1469$

b. $P(x < 6) = P(z < \frac{6 - 8}{1.9}) = P(z < -1.05) = .5 - .3531 = .1469$

Define the following events:

A: First cigarette has less than 6 milligrams of tar
B: Second cigarette has less than 6 milligrams of tar

$$P(A) = P(B) = .1469$$

The event "both cigarettes have less than 6 milligrams of tar" is AB.

$$P(AB) = P(A)P(B) = .1469(.1469) = .0216$$
(assume A and B are independent)

4.121 Let x = number of American blacks with sickle-cell anemia. Then x is a binomial random variable with n = 1000 and p = .16.

$$\mu = np = 1000(.16) = 160 \text{ and } \sigma = \sqrt{npq} = \sqrt{1000(.16)(.84)} = 11.593$$

Using the normal approximation to the binomial and Table III,

a. $P(x > 175) \approx P(z > \frac{175 + .5 - 160}{11.593}) = P(z > 1.34) = .5 - .4099$
$= .0901$

b. $P(x < 140) \approx P(z < \frac{140 - .5 - 160}{11.593}) = P(z < -1.77) = .5 - .4616$
$= .0384$

c. $P(130 \leq x \leq 180) \approx P(\frac{130 - .5 - 160}{11.593} \leq z \leq \frac{180 + .5 - 160}{11.593})$
$= P(-2.63 \leq z \leq 1.77) = .4957 + .4616$
$= .9573$

4.123 a. $E(x) = \sum xp(x) = 37{,}500(.14) + 112{,}500(.30) + 225{,}000(.20) + 400{,}000(.17) + 750{,}000(.10) + 3{,}000{,}000(.09)$
$= \$497{,}000$

b. $E[(x - \mu)^2] = \sum(x - \mu)^2 p(x)$

$= (37{,}500 - 497{,}000)^2(.14) + (112{,}500 - 497{,}000)^2(.30) + (225{,}000 - 497{,}000)^2(.20) + (400{,}000 - 497{,}000)^2(.17) + (750{,}000 - 497{,}000)^2(.10) + (3{,}000{,}000 - 497{,}000)^2(.09)$

$= 6.6056 \times 10^{11}$

c. The average sales volumes for fabricare firms is $\mu = \$497{,}000$.

4.125 Let x = sales per salesperson. Using Table III, Appendix A:

a. $p_1 = P(x < 100{,}000) = P(z < \frac{100{,}000 - 180{,}000}{50{,}000}) = P(z < -1.6)$
$= .5 - .4452$
$= .0548$

b. $p_2 = P(100{,}000 \leq x \leq 200{,}000)$
$= P(\frac{100{,}000 - 180{,}000}{50{,}000} < z < \frac{200{,}000 - 180{,}000}{50{,}000})$
$= P(-1.60 < z < .40) = .4452 + .1554 = .6006$

c. $p_3 = P(x > 200{,}000) = P(z > \frac{200{,}000 - 180{,}000}{50{,}000}) = P(x > .40)$
$= .5 - .1554$
$= .3446$

d. Let y = bonus payout. The probability distribution for y is

y	p(y)
\$ 1,000	.0548
\$ 5,000	.6006
\$10,000	.3446

$$\mu = E(y) = \sum yp(y) = 1000(.0548) + 5000(.6006) + 10,000(.3446)$$
$$= 54.8 + 3003 + 3446$$
$$= \$6503.80$$

4.127 Let x = the number that showed improvement in 15 trials. Then x is a binomial random variable with $n = 15$ and $p = .5$.

$P(x \geq 11) = 1 - P(x \leq 10) = 1 - .941 = .059$ (Table II, Appendix A)

Since we usually do not "see" rare events (probability .059), we would conclude the probability of improvement exceeds $p = 1/2$ and thus, the program is effective.

4.129 For $n = 1600$ and $p = .2$, $\mu = np = 1600(.2) = 320$ and

$\sigma = \sqrt{npq} = \sqrt{1600(.2)(.8)} = 16$

Using the normal approximation to the binomial and Table III,

$$P(x \geq 400) \approx P(z \geq \frac{400 - .5 - 320}{16}) = P(z \geq 4.97) = .5 - .5 = 0$$

If $p = .2$, the probability of observing 400 or more consumers who favor the product is essentially 0. This implies that p is probably not .2 but larger than .2.

CHAPTER 5

SAMPLING DISTRIBUTIONS

5.1 a. - b. The different samples of $n = 2$ and their means are:

POSSIBLE SAMPLES	$\bar{x}$	POSSIBLE SAMPLES	$\bar{x}$
0, 0	0	4, 0	2
0, 2	1	4, 2	3
0, 4	2	4, 4	4
0, 6	3	4, 6	5
2, 0	1	6, 0	3
2, 2	2	6, 2	4
2, 4	3	6, 4	5
2, 6	4	6, 6	6

c. Since each sample is equally likely, the probability of any 1 being selected is $\frac{1}{4}\left(\frac{1}{4}\right) = \frac{1}{16}$

d. $P(\bar{x} = 0) = \frac{1}{16}$

$P(\bar{x} = 1) = \frac{1}{16} + \frac{1}{16} = \frac{2}{16}$

$P(\bar{x} = 2) = \frac{1}{16} + \frac{1}{16} + \frac{1}{16} = \frac{3}{16}$

$P(\bar{x} = 3) = \frac{1}{16} + \frac{1}{16} + \frac{1}{16} + \frac{1}{16} = \frac{4}{16}$

$P(\bar{x} = 4) = \frac{1}{16} + \frac{1}{16} + \frac{1}{16} = \frac{3}{16}$

$P(\bar{x} = 5) = \frac{1}{16} + \frac{1}{16} = \frac{2}{16}$

$P(\bar{x} = 6) = \frac{1}{16}$

$\bar{x}$	$p(\bar{x})$
0	1/16
1	2/16
2	3/16
3	4/16
4	3/16
5	2/16
6	1/16

e.

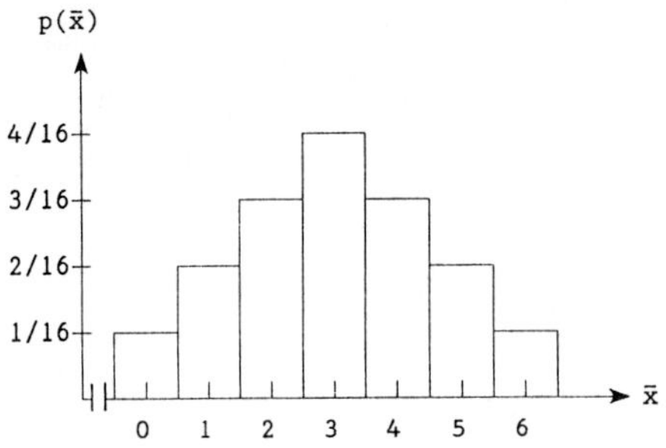

5.3 If the observations are independent of each other, then

$P(1, 1) = p(1)p(1) = .2(.2) = .04$
$P(1, 2) = p(1)p(2) = .2(.3) = .06$
$P(1, 3) = p(1)p(3) = .2(.2) = .04$
etc.

a.

POSSIBLE SAMPLES	$\bar{x}$	$p(\bar{x})$	POSSIBLE SAMPLES	$\bar{x}$	$p(\bar{x})$
1, 1	1	.04	3, 4	3.5	.04
1, 2	1.5	.06	3, 5	4	.02
1, 3	2	.04	4, 1	2.5	.04
1, 4	2.5	.04	4, 2	3	.06
1, 5	3	.02	4, 3	3.5	.04
2, 1	1.5	.06	4, 4	4	.04
2, 2	2	.09	4, 5	4.5	.02
2, 3	2.5	.06	5, 1	3	.02
2, 4	3	.06	5, 2	3.5	.03
2, 5	3.5	.03	5, 3	4	.02
3, 1	2	.04	5, 4	4.5	.02
3, 2	2.5	.06	5, 5	5	.01
3, 3	3	.04			

Summing the probabilities, the probability distribution of $\bar{x}$ is:

$\bar{x}$	$p(\bar{x})$
1	.04
1.5	.12
2	.17
2.5	.20
3	.20
3.5	.14
4	.08
4.5	.04
5	.01

b.

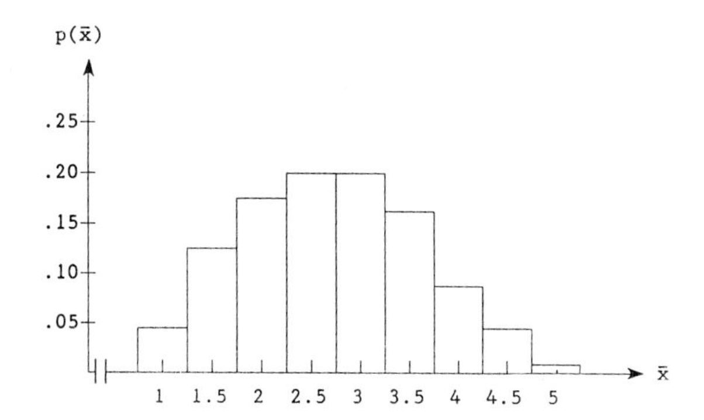

c. $P(\bar{x} \geq 4.5) = .04 + .01 = .05$

d. No. The probability of observing $\bar{x} = 4.5$ or larger is small (.05).

5.5 a. For a sample of size n = 2, the sample mean and sample median are exactly the same. Thus, the sampling distribution of the sample median is the same as that for the sample mean (see Exercise 5.3(a)).

b. The probability histogram for the sample median is identical to that for the sample mean (see Exercise 5.3(b)).

5.7 a. The possible samples, sample medians, and probabilities are:

POSSIBLE SAMPLES	m	p(m)	POSSIBLE SAMPLES	m	p(m)
1, 1, 1	1	.216	2, 3, 2	2	.009
1, 1, 2	1	.108	2, 1, 3	2	.018
1, 2, 1	1	.108	2, 3, 1	2	.018
1, 2, 2	2	.054	2, 3, 3	3	.003
1, 2, 3	2	.018	3, 1, 1	1	.036
1, 3, 2	2	.018	3, 1, 2	2	.018
1, 1, 3	1	.036	3, 2, 1	2	.018
1, 3, 1	1	.036	3, 2, 2	2	.009
1, 3, 3	3	.006	3, 2, 3	3	.003
2, 1, 1	1	.108	3, 3, 2	3	.003
2, 1, 2	2	.054	3, 1, 3	3	.006
2, 2, 1	2	.054	3, 3, 1	3	.006
2, 2, 2	2	.027	3, 3, 3	3	.001
2, 2, 3	2	.009			

The probabilities, assuming the observations are independent, are

$P(1, 1, 1) = p(1)p(1)p(1) = .6(.6)(.6) = .216$

$P(1, 1, 2) = p(1)p(1)p(2) = .108$ etc.

Summing the probabilities, the sampling distribution of m is:

m	p(m)
1	.648
2	.324
3	.028

b.

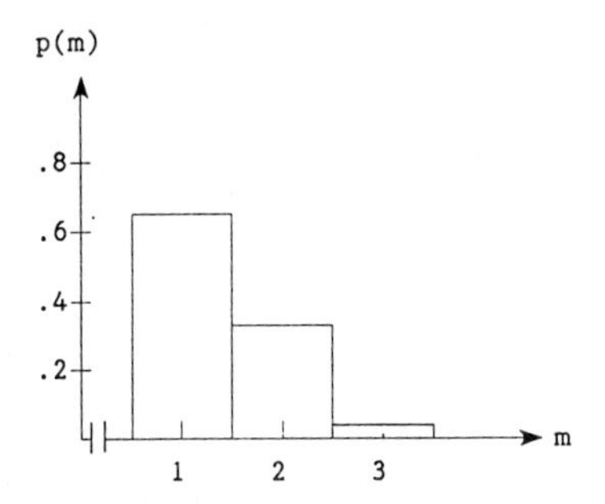

The basic shapes of the 2 probability distributions are similar--skewed to the right.

5.9 a. The possible samples if size n = 4, the sample medians, and their probabilities are:

POSSIBLE SAMPLES	m	p(m)	POSSIBLE SAMPLES	m	p(m)	POSSIBLE SAMPLES	m	p(m)
1,1,1,1	1	.1296	2,1,1,1	1	.0648	3,1,1,1	1	.0216
1,1,1,2	1	.0648	2,1,1,2	1.5	.0324	3,1,1,2	1.5	.0108
1,1,1,3	1	.0216	2,1,1,3	1.5	.0108	3,1,1,3	2	.0036
1,1,2,1	1	.0648	2,1,2,1	1.5	.0324	3,1,2,1	1.5	.0108
1,1,2,2	1.5	.0324	2,1,2,2	2	.0162	3,1,2,2	2	.0054
1,1,2,3	1.5	.0108	2,1,2,3	2	.0054	3,1,2,3	2.5	.0018
1,1,3,1	1	.0216	2,1,3,1	1.5	.0108	3,1,3,1	2	.0036
1,1,3,2	1.5	.0108	2,1,3,2	2	.0054	3,1,3,2	2.5	.0018
1,1,3,3	2	.0036	2,1,3,3	2.5	.0018	3,1,3,3	3	.0006
1,2,1,1	1	.0648	2,2,1,1	1.5	.0324	3,2,1,1	1.5	.0108
1,2,1,2	1.5	.0324	2,2,1,2	2	.0162	3,2,1,2	2	.0054
1,2,1,3	1.5	.0108	2,2,1,3	2	.0054	3,2,1,3	2.5	.0018
1,2,2,1	1.5	.0324	2,2,2,1	2	.0162	3,2,2,1	2	.0054
1,2,2,2	2	.0162	2,2,2,2	2	.0081	3,2,2,2	2	.0027
1,2,2,3	2	.0054	2,2,2,3	2	.0027	3,2,2,3	2.5	.0009
1,2,3,1	1.5	.0108	2,2,3,1	2	.0054	3,2,3,1	2.5	.0018
1,2,3,2	2	.0054	2,2,3,2	2	.0027	3,2,3,2	2.5	.0009
1,2,3,3	2.5	.0018	2,2,3,3	2.5	.0009	3,2,3,3	3	.0003
1,3,1,1	1	.0216	2,3,1,1	1.5	.0108	3,3,1,1	2	.0036
1,3,1,2	1.5	.0108	2,3,1,2	2	.0054	3,3,1,2	2.5	.0018
1,3,1,3	2	.0036	2,3,1,3	2.5	.0018	3,3,1,3	3	.0006
1,3,2,1	1.5	.0108	2,3,2,1	2	.0054	3,3,2,1	2.5	.0018
1,3,2,2	2	.0054	2,3,2,2	2	.0027	3,3,2,2	2.5	.0009
1,3,2,3	2.5	.0018	2,3,2,3	2.5	.0009	3,3,2,3	3	.0003
1,3,3,1	2	.0036	2,3,3,1	2.5	.0018	3,3,3,1	3	.0006
1,3,3,2	2.5	.0018	2,3,3,2	2.5	.0009	3,3,3,2	3	.0003
1,3,3,3	3	.0006	2,3,3,3	3	.0003	3,3,3,3	3	.0001

b. The sampling distribution of the sample median is:

m	p(m)
1	.4752
1.5	.3240
2	.1701
2.5	.0270
3	.0037
	1.0000

c.

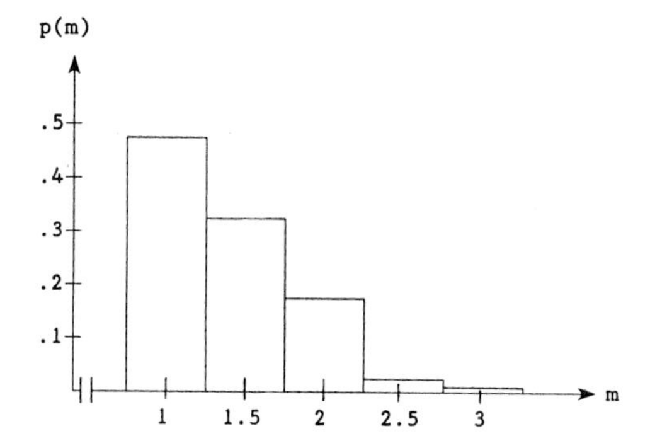

As the sample size increases, the distribution of the sample median is shifting somewhat to the left.

5.13 a. $\mu = \sum xp(x) = 2\left(\frac{1}{3}\right) + 4\left(\frac{1}{3}\right) + 9\left(\frac{1}{3}\right) = \frac{15}{3} = 5$

b. The possible samples of size n = 3, the sample means, the probabilities, and medians are:

POSSIBLE SAMPLES	$\bar{x}$	$p(\bar{x})$	m	POSSIBLE SAMPLES	$\bar{x}$	$p(\bar{x})$	m
2, 2, 2	2	1/27	2	4, 4, 4	4	1/27	4
2, 2, 4	8/3	1/27	2	4, 4, 9	17/3	1/27	4
2, 2, 9	13/3	1/27	2	4, 9, 2	5	1/27	4
2, 4, 2	8/3	1/27	2	4, 9, 4	17/3	1/27	4
2, 4, 4	10/3	1/27	4	4, 9, 9	22/3	1/27	9
2, 4, 9	5	1/27	4	9, 2, 2	13/3	1/27	2
2, 9, 2	13/3	1/27	2	9, 2, 4	5	1/27	4
2, 9, 4	5	1/27	4	9, 2, 9	20/3	1/27	9
2, 9, 9	20/3	1/27	9	9, 4, 2	5	1/27	4
4, 2, 2	8/3	1/27	2	9, 4, 4	17/3	1/27	4
4, 2, 4	10/3	1/27	4	9, 4, 9	22/3	1/27	9
4, 2, 9	5	1/27	4	9, 9, 2	20/3	1/27	9
4, 4, 2	10/3	1/27	4	9, 9, 4	22/3	1/27	9
				9, 9, 9	9	1/27	9

The sampling distribution of $\bar{x}$ is:

$\bar{x}$	$p(\bar{x})$
2	1/27
8/3	3/27
10/3	3/27
4	1/27
13/3	3/27
5	6/27
17/3	3/27
20/3	3/27
22/3	3/27
9	1/27
	27/27

$$E(\bar{x}) = \sum \bar{x}p(\bar{x}) = 2\left(\frac{1}{27}\right) + \frac{8}{3}\left(\frac{3}{27}\right) + \frac{10}{3}\left(\frac{3}{27}\right) + 4\left(\frac{1}{27}\right) + \frac{13}{3}\left(\frac{3}{27}\right)$$

$$+ 5\left(\frac{6}{27}\right) + \frac{17}{3}\left(\frac{3}{27}\right) + \frac{20}{3}\left(\frac{3}{27}\right) + \frac{22}{3}\left(\frac{3}{27}\right) + 9\left(\frac{1}{27}\right)$$

$$= \frac{2}{27} + \frac{8}{27} + \frac{10}{27} + \frac{4}{27} + \frac{13}{27} + \frac{30}{27} + \frac{17}{27} + \frac{20}{27} + \frac{22}{27} + \frac{9}{27}$$

$$= \frac{135}{27} = 5$$

Since $\mu = 5$ in part (a), and $E(\bar{x}) = \mu = 5$, $\bar{x}$ is an unbiased estimator of μ.

c. The sampling distribution of m is

m	p(m)
2	7/27
4	13/27
9	7/27
	27/27

$$E(m) = \sum mp(m) = 2\left(\frac{7}{27}\right) + 4\left(\frac{13}{27}\right) + 9\left(\frac{7}{27}\right)$$

$$= \frac{14}{27} + \frac{52}{27} + \frac{63}{27}$$

$$= \frac{129}{27} = 4.778$$

The $E(m) = 4.778 \neq \mu = 5$. Thus, m is a biased estimator of μ.

d. Use the sample mean, $\bar{x}$. It is an unbiased estimator.

5.17 a. Refer to the solution to Exercise 5.7. The values of s^2 and the corresponding probabilities are listed below:

$$s^2 = \frac{\sum(x^2) - \frac{(\sum x)^2}{n}}{n - 1}$$

For sample 1, 1, 1, $s^2 = \dfrac{3 - \frac{3^2}{3}}{2} = 0$

For sample 1, 1, 2, $s^2 = \dfrac{6 - \frac{4^2}{3}}{2} = .333$

The rest of the values are calculated and shown:

s^2	$p(s^2)$	s^2	$p(s^2)$
0	.216	.333	.009
.333	.108	1	.018
.333	.108	1	.018
.333	.054	.333	.003
1	.018	1.333	.036
1	.018	1	.018
1.333	.036	1	.018
1.333	.036	.333	.009
1.333	.006	.333	.003
.333	.108	.333	.003
.333	.054	1.333	.006
.333	.054	1.333	.006
0	.027	0	.001
.333	.009		

The sampling distribution of s^2 is:

s^2	$p(s^2)$
0	.244
.333	.522
1	.108
1.333	.126
	1.000

b. $\sigma^2 = \sum(s^2 - \mu_{s^2})^2 p(s^2) = (1 - 1.5)^2(.6) + (2 - 1.5)^2(.3) + (3 - 1.5)^2(.1)$
$= .15 + .075 + .225$
$= .45$

c. $E(s^2) = \sum s^2 p(s^2) = 0(.244) + .333(.522) + 1(.108) + 1.333(.126)$
$= 0 + .174 + .108 + .168 = .45$

d. The sampling distribution of s is listed below, where $s = \sqrt{s^2}$:

s	p(s)
0	.244
.577	.522
1	.108
1.155	.126

e. $E(s) = \sum sp(s) = 0(.244) + .577(.522) + 1(.108) + 1.155(.126)$
$= 0 + .301 + .108 + .146$
$= .555$

Since $E(s) = .555$ is not equal to $\sigma = \sqrt{\sigma^2} = \sqrt{.45} = .671$, s is a biased estimator of σ.

5.19 a. $\mu_{\bar{x}} = \mu = 50$, $\sigma_{\bar{x}} = \frac{\sigma}{\sqrt{n}} = \frac{\sqrt{50}}{\sqrt{4}} = 3.536$

b. $\mu_{\bar{x}} = \mu = 50$, $\sigma_{\bar{x}} = \frac{\sigma}{\sqrt{n}} = \frac{\sqrt{50}}{\sqrt{25}} = 1.414$

c. $\mu_{\bar{x}} = \mu = 50$, $\sigma_{\bar{x}} = \frac{\sigma}{\sqrt{n}} = \frac{\sqrt{50}}{\sqrt{100}} = .707$

d. $\mu_{\bar{x}} = \mu = 50$, $\sigma_{\bar{x}} = \frac{\sigma}{\sqrt{n}} = \frac{\sqrt{50}}{\sqrt{50}} = 1$

e. $\mu_{\bar{x}} = \mu = 50$, $\sigma_{\bar{x}} = \frac{\sigma}{\sqrt{n}} = \frac{\sqrt{50}}{\sqrt{500}} = .316$

f. $\mu_{\bar{x}} = \mu = 50$, $\sigma_{\bar{x}} = \frac{\sigma}{\sqrt{n}} = \frac{\sqrt{50}}{\sqrt{1000}} = .224$

5.21 a. $\mu = \sum xp(x) = 1(.1) + 2(.4) + 3(.4) + 10(.1) = 3.1$

$\sigma^2 = \sum(x - \mu)^2p(x) = (1 - 3.1)^2(.1) + (2 - 3.1)^2(.4)$
$+ (3 - 3.1)^2(.4) + (10 - 3.1)^2(.1)$
$= .441 + .484 + .004 + 4.761$
$= 5.69$

$\sigma = \sqrt{5.69} = 2.385$

b. The possible samples, values of $\bar{x}$, and associated probabilities are listed:

b. Yes. By the Central Limit Theorem, the distribution of $\bar{x}$ is approximately normal for n sufficiently large. For this problem, n = 50 is sufficiently large.

c. $P(\bar{x} < 1) = P(z < \frac{1 - 1.3}{.24}) = P(z < -1.25) = .5 - .3944 = .1056$

(Using Table III, Appendix A)

d. $P(\bar{x} > 1.9) = P(z > \frac{1.9 - 1.3}{.24}) = P(z > 2.5) = .5 - .4938 = .0062$

(Using Table III, Appendix A)

5.33 a. By the Central Limit Theorem, the distribution of $\bar{x}$ is approximately normal, with $\mu_{\bar{x}} = \mu = 157$ and $\sigma_{\bar{x}} = \sigma/\sqrt{n} = 3/\sqrt{40} = .474$.

The sample mean is 1.3 psi below 157 or $\bar{x} = 157 - 1.3 = 155.7$

$P(\bar{x} \leq 155.7) = P(z \leq \frac{155.7 - 157}{.474}) = P(z \leq -2.74) = .5 - .4969 = .0031$

(Using Table III, Appendix A)

If the claim is true, it is very unlikely (probability = .0031) to observe a sample mean 1.3 psi below 157 psi. Thus, the actual population mean is probably not 157 but something lower.

b. $P(\bar{x} \leq 155.7) = P(z \leq \frac{155.7 - 156}{.474}) = P(z \leq -.63) = .5 - .2357 = .2643$

(Using Table III, Appendix A)

The observed sample is more likely if $\mu = 156$ rather than 157.

$P(\bar{x} \leq 155.7) = P(z \leq \frac{155.7 - 158}{.474}) = P(z \leq -4.85) = .5 - .5 = 0$

The observed sample is less likely if $\mu = 158$ rather than 157.

c. $\sigma_{\bar{x}} = \sigma/\sqrt{n} = 2/\sqrt{40} = .316 \quad \mu_{\bar{x}} = 157$

$P(\bar{x} \leq 155.7) = P(z \leq \frac{155.7 - 157}{.316}) = P(z \leq -4.11) = .5 - .5 = 0$

(Using Table III, Appendix A)

The observed sample is less likely if $\sigma = 2$ rather than 3.

$\sigma_{\bar{x}} = \sigma/\sqrt{n} = 6/\sqrt{40} = .949 \quad \mu_{\bar{x}} = 157$

$$P(\bar{x} \leq 155.7) = P\left(z \leq \frac{155.7 - 157}{.949}\right) = P(z \leq -1.37) = .5 - .4147$$
$$= .0853$$

(Using Table III, Appendix A)

The observed sample is more likely if $\sigma = 6$ rather than 3.

5.35 a. The population of interest is the market value of all single-family homes in a particular county in 1987.

b. By the Central Limit Theorem, the sampling distribution of $\bar{x}$ is approximately normal. The probability of observing a value higher than μ is .5.

c. Having the sample mean fall within \$4000 of μ implies $|\bar{x} - \mu| \leq 4000$ or $-4000 \leq \bar{x} - \mu \leq 4000$.

$$P(-4000 \leq \bar{x} - \mu \leq 4000) = P\left(\frac{-4000}{\sigma_{\bar{x}}} \leq z \leq \frac{4000}{\sigma_{\bar{x}}}\right)$$

$$= P\left(\frac{-4000}{\frac{50,000}{\sqrt{400}}} \leq z \leq \frac{4000}{\frac{50,000}{\sqrt{400}}}\right) = P(-1.60 \leq z \leq 1.60)$$

$$= 2P(0 \leq z \leq 1.60) = 2(.4452) = .8904$$

5.37 a. As the sample size increases, the standard error will decrease. This property is important because we know that the larger the sample size, the less variable our estimator will be. Thus, as n increases, our estimator will tend to be closer to the parameter we are trying to estimate.

b. This would indicate that the statistic would not be a very good estimator of the parameter. If the standard error is not a function of the sample size, then a statistic based on one observation would be as good an estimator as a statistic based on 1000 observations.

c. $\bar{x}$ would be preferred over A as an estimator for the population mean. The standard error of $\bar{x}$ is smaller than the standard error of A.

d. The standard error of $\bar{x}$ is $\sigma/\sqrt{n} = 10/\sqrt{64} = 1.25$ and the standard error of A is $\sigma/\sqrt[3]{64} = 2.5$.

If the sample size is sufficiently large, the Central Limit Theorem says the distribution of $\bar{x}$ is approximately normal. Using the Empirical Rule, approximately 68% of all the values of $\bar{x}$ will fall between $\mu - 1.25$ and $\mu + 1.25$. Approximately 95% of all the values of $\bar{x}$ will fall between $\mu - 2.50$ and $\mu + 2.50$. Approximately all of the values of $\bar{x}$ will fall between $\mu - 3.75$ and $\mu + 3.75$.

Using the Empirical Rule, approximately 68% of all the values of A will fall between $\mu - 2.50$ and $\mu + 2.50$. Approximately 95% of all the values of A will fall between $\mu - 5.00$ and $\mu + 5.00$. Approximately all of the values of A will fall between $\mu - 7.50$ and $\mu + 7.50$.

5.39 a. First we must compute μ and σ. The probability distribution for x is:

x	p(x)
1	.3
2	.2
3	.2
4	.3

$$\mu = E(x) = \sum xp(x) = 1(.3) + 2(.2) + 3(.2) + 4(.3) = 2.5$$

$$\sigma^2 = E(x^2) - \mu^2 = 1^2(.3) + 2^2(.2) + 3^2(.2) + 4^2(.3) - 2.5^2 = 1.45$$

$$\mu_{\bar{x}} = \mu = 2.5, \quad \sigma_{\bar{x}} = \frac{\sigma}{\sqrt{n}} = \frac{\sqrt{1.45}}{\sqrt{40}} = .1904$$

b. By the Central Limit Theorem, the distribution of $\bar{x}$ is approximately normal. The sample size, $n = 40$, is sufficiently large.

5.41 a. $\mu_{\bar{x}} = \mu = 120, \quad \sigma_{\bar{x}} = \frac{\sigma}{\sqrt{n}} = \frac{\sqrt{410}}{\sqrt{75}} = 2.338$

b. By the Central Limit Theorem, the distribution of $\bar{x}$ is approximately normal. The sample size, $n = 75$, is sufficiently large. The Central Limit Theorem holds regardless of the population shape.

c. $P(\bar{x} \leq 118) = P(z \leq \frac{118 - 120}{2.338}) = P(z \leq -.86) = .5 - .3051 = .1949$

(Using Table III, Appendix A)

d. $$P(115 \leq \bar{x} \leq 123) = P(\frac{115 - 120}{2.338} \leq z \leq \frac{123 - 120}{2.338})$$
$$= P(-2.14 \leq z \leq 1.28)$$
$$= .4838 + .3997 = .8835$$

e. $P(\bar{x} \leq 124.6) = P(z \leq \frac{124.6 - 120}{2.338}) = P(z \leq 1.97) = .5 + .4756$
$= .9756$

f. $P(\bar{x} \geq 122.7) = P(z \geq \frac{122.7 - 120}{2.338}) = P(z \geq 1.15) = .5 - .3749$
$= .1251$

5.43 a. The sample space, sample mean, and probabilities are listed:

POSSIBLE SAMPLES	$\bar{x}$	$p(\bar{x})$	POSSIBLE SAMPLES	$\bar{x}$	$p(\bar{x})$	POSSIBLE SAMPLES	$\bar{x}$	$p(\bar{x})$
1, 1	1	1/36	3, 1	2	1/36	5, 1	3	1/36
1, 2	1.5	1/36	3, 2	2.5	1/36	5, 2	3.5	1/36
1, 3	2	1/36	3, 3	3	1/36	5, 3	4	1/36
1, 4	2.5	1/36	3, 4	3.5	1/36	5, 4	4.5	1/36
1, 5	3	1/36	3, 5	4	1/36	5, 5	5	1/36
1, 6	3.5	1/36	3, 6	4.5	1/36	5, 6	5.5	1/36
2, 1	1.5	1/36	4, 1	2.5	1/36	6, 1	3.5	1/36
2, 2	2	1/36	4, 2	3	1/36	6, 2	4	1/36
2, 3	2.5	1/36	4, 3	3.5	1/36	6, 3	4.5	1/36
2, 4	3	1/36	4, 4	4	1/36	6, 4	5	1/36
2, 5	3.5	1/36	4, 5	4.5	1/36	6, 5	5.5	1/36
2, 6	4	1/36	4, 6	5	1/36	6, 6	6	1/36

The sampling distribution of $\bar{x}$ is:

$\bar{x}$	$p(\bar{x})$
1	1/36
1.5	2/36
2	3/36
2.5	4/36
3	5/36
3.5	6/36
4	5/36
4.5	4/36
5	3/36
5.5	2/36
6	1/36

b. $P(\bar{x} \leq 5) = 1 - P(\bar{x} = 5.5) - P(\bar{x} = 6) = 1 - \frac{2}{36} - \frac{1}{36} = \frac{33}{36}$

5.45 a. Tossing a coin two times can result in:

2 heads (2 ones)
2 tails (2 zeros)
1 head, 1 tail (1 one, 1 zero)

b. $\bar{x}_{2\ \text{heads}} = 1$; $\bar{x}_{2\ \text{tails}} = 0$; $\bar{x}_{1H,\ 1T} = \frac{1}{2}$

c. There are four possible combinations for one coin tossed two times, as shown below:

Coin Tosses	$\bar{x}$	$\bar{x}$	$p(\bar{x})$
H, H	1	0	1/4
H, T	1/2	1/2	1/2
T, H	1/2	1	1/4
T, T	0		

d. The sampling distribution of $\bar{x}$ is given in the histogram below:

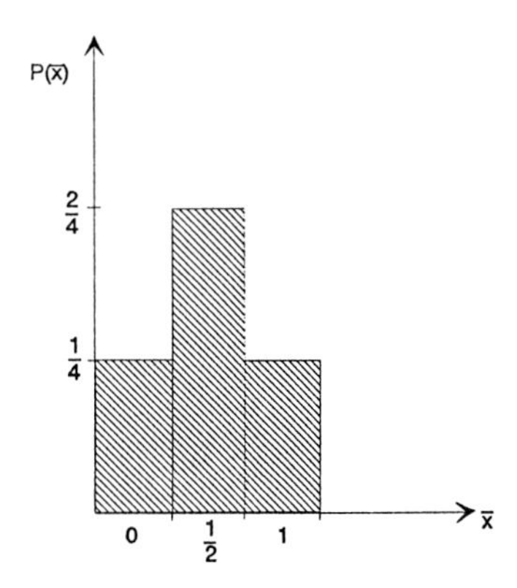

5.49 Given: $\mu = 100$ and $\sigma = 10$

n	1	5	10	20	30	40	50
$\frac{\sigma}{\sqrt{n}}$	10	4.472	3.162	2.236	1.826	1.581	1.414

The graph of $\sigma/\sqrt{n}$ against n is given below:

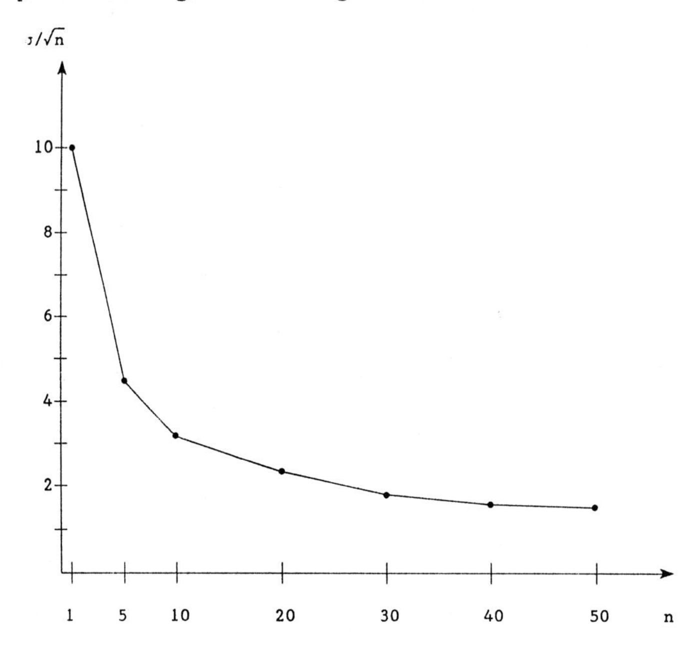

5.51 $\mu_{\bar{x}} = \mu = 8.9, \quad \sigma_{\bar{x}} = \frac{.13}{\sqrt{35}} = .022$

a. $P(\bar{x} < 8.95) = P(z < \frac{8.95 - 8.9}{.022}) = P(z < 2.28) = .5 + .4887$

$= .9887$ (from Table III, Appendix A)

b. The assumption is that the distribution of the sample mean is approximately normal. Based on the Central Limit Theorem, this assumption is reasonable.

5.53 $\mu_{\bar{x}} = \mu = 3.1, \quad \sigma_{\bar{x}} = \frac{.8}{\sqrt{34}} = .137$

a. $P(\bar{x} < 3) = P(z < \frac{3 - 3.1}{.137}) = P(z < -0.73) = .5 - .2673 = .2327$

(from Table III, Appendix A)

b. $P(3.15 < \bar{x} < 3.25) = P(\frac{3.15 - 3.1}{.137} < z < \frac{3.25 - 3.1}{.137})$

$= P(.36 < z < 1.09) = .3621 + .1406 = .2215$

5.55 a. The mean, μ, diameter of the bearings is unknown with a standard deviation, σ, of .001 inch. Assuming that the distribution of the diameters of the bearings is normal, the sampling distribution of the sample mean is also normal. The mean and variance of the distribution are:

$$\mu_{\bar{x}} = \mu \qquad \sigma_{\bar{x}} = \frac{\sigma}{\sqrt{n}} = \frac{.001}{\sqrt{25}} = .0002$$

Having the sample mean fall within .0001 inch of μ implies

$$|\bar{x} - \mu| \leq .0001 \text{ or } -.0001 \leq \bar{x} - \mu \leq .0001$$

$$P(-.0001 \leq \bar{x} - \mu \leq .0001)$$

$$= P\left(\frac{-.0001}{.0002} \leq z \leq \frac{.0001}{.0002}\right)$$

$$= P(-.50 \leq z \leq .50)$$

$$= 2P(0 \leq z \leq .50) = 2(.1915) = .3830$$

b. Since the Central Limit Theorem only applies to the large sample case, the accuracy of the above calculations would be brought into question. A skewed distribution would mean that the mean of the distribution would be pulled toward the tail and away from the bulk of the distribution. Thus, the probability would decrease.

5.57 a. There are $\binom{10}{2} = \frac{10!}{2!8!} = 45$ possible samples. They are:

1,2	1,3	1,4	1,5	1,6	1,7	1,8	1,9	1,10
2,3	2,4	2,5	2,6	2,7	2,8	2,9	2,10	3,4
3,5	3,6	3,7	3,8	3,9	3,10	4,5	4,6	4,7
4,8	4,9	4,10	5,6	5,7	5,8	5,9	5,10	6,7
6,8	6,9	6,10	7,8	7,9	7,10	8,9	8,10	9,10

b. The samples which contain two good bottles are:

1,4	1,5	1,6	1,7	1,8	1,10	4,5
4,6	4,7	4,8	4,10	5,6	5,7	5,8
5,10	6,7	6,8	6,10	7,8	7,10	8,10

Since each possible sample is equally likely, the probability of accepting the case is equal to the number of samples containing two good bottles (21), divided by the number of possible samples (45), i.e., 21/45.

c. Among the possible samples, only (2,3), (2,9), and (3,9) have zero good bottles. This, plus the result of part (b), imply $p(0) = 3/45$, $p(1) = 21/45$, $p(2) = 21/45$.

5.59 a. $\mu_{\bar{x}} = \mu = 42$, $\sigma_{\bar{x}} = \dfrac{6.5}{\sqrt{100}} = .65$

In addition to the values of the parameters above, since the sample size is large, the distribution of the sample mean will be approximately normal by the Central Limit Theorem.

b. $P(\bar{x} > 43.6) = P(z > \dfrac{43.6 - 42}{.65}) = P(z > 2.46) = .5 - .4931 = .0069$

(from Table III, Appendix A)

c. Since the probability of obtaining this sample mean is only .0069, if this were a sample from the general public, the logical conclusion is that the health fair attendees are not a random sample of the general public with regard to the malignancy questionnaire.

5.61 a. $P(\bar{x} > 79) = P(z > \dfrac{79 - 75}{10/\sqrt{36}}) = P(z > 2.4) = .5 - .4918 = .0082$

(from Table III, Appendix A)

b. Since the sample size exceeds 30, the Central Limit Theorem implies that the distribution of the sample mean will be approximately normal, so that no assumptions are necessary.

CHAPTER 6

INFERENCES BASED ON A SINGLE SAMPLE: ESTIMATION

6.1 a. For $\alpha = .10$, $\alpha/2 = .10/2 = .05$. $z_{\alpha/2} = z_{.05}$ is the z-score with .05 of the area to the right of it. The area between 0 and $z_{.05}$ is $.5 - .05 = .4500$. Using Table III, Appendix A, $z_{.05} = 1.645$.

b. For $\alpha = .01$, $\alpha/2 = .01/2 = .005$. $z_{\alpha/2} = z_{.005}$ is the z-score with .005 of the area to the right of it. The area between 0 and $z_{.005}$ is $.5 - .005 = .4950$. Using Table III, Appendix A, $z_{.005} = 2.575$.

c. For $\alpha = .05$, $\alpha/2 = .05/2 = .025$. $z_{\alpha/2} = z_{.025}$ is the z-score with .025 of the area to the right of it. The area between 0 and $z_{.025}$ is $.5 - .025 = .4750$. Using Table III, Appendix A, $z_{.025} = 1.96$.

d. For $\alpha = .20$, $\alpha/2 = .20/2 = .10$. $z_{\alpha/2} = z_{.10}$ is the z-score with .10 of the area to the right of it. The area between 0 and $z_{.10}$ is $.5 - .10 = .4000$. Using Table III, Appendix A, $z_{.10} = 1.28$.

6.3 a. $s^2 = \dfrac{\sum(x_i - \bar{x})^2}{n - 1} = \dfrac{3566}{64 - 1} = 56.6032$, $s = \sqrt{56.6032} = 7.5235$

$\bar{x} = \dfrac{\sum x}{n} = \dfrac{500}{64} = 7.8125$

For confidence coefficient .95, $\alpha = .05$ and $\alpha/2 = .05/2 = .025$. From Table III, Appendix A, $z_{.025} = 1.96$. The confidence interval is:

$$\bar{x} \pm z_{\alpha/2} \frac{s}{\sqrt{n}} \Rightarrow 7.8125 \pm 1.96 \frac{7.5235}{\sqrt{64}} \Rightarrow 7.8125 \pm 1.8433$$

$$\Rightarrow (5.9692,\ 9.6558)$$

b. We are 95% confident the true mean is between 5.9692 and 9.6558.

6.5 a. For confidence coefficient .95, $\alpha = .05$ and $\alpha/2 = .05/2 = .025$. From Table III, Appendix A, $z_{.025} = 1.96$. The confidence interval is:

$$\bar{x} \pm z_{\alpha/2} \frac{s}{\sqrt{n}} \Rightarrow 76.9 \pm 1.96 \frac{4.8}{\sqrt{100}} \Rightarrow 76.9 \pm .94 \Rightarrow (75.96,\ 77.84)$$

b. The confidence coefficient of .95 means that in repeated sampling, 95% of all confidence intervals constructed will include μ.

c. For confidence coefficient .99, $\alpha = .01$ and $\alpha/2 = .01/2 = .005$. From Table III, Appendix A, $z_{.005} = 2.575$. The confidence interval is:

$$\bar{x} \pm z_{\alpha/2} \frac{s}{\sqrt{n}} \Rightarrow 76.9 \pm 2.575 \frac{4.8}{\sqrt{100}} \Rightarrow 76.9 \pm 1.24 \Rightarrow (75.66, 78.14)$$

d. As the confidence coefficient increases, the width of the confidence interval also increases.

e. Yes. Since the sample size is 100, the Central Limit Theorem applies. This ensures the distribution of $\bar{x}$ is normal, regardless of the original distribution.

6.7 An interval estimator estimates μ with a range of values while a point estimator estimates μ with a single point.

6.9 Yes. As long as the sample size is sufficiently large, the Central Limit Theorem says the distribution of $\bar{x}$ is approximately normal regardless of the original distribution.

6.11 For confidence coefficient .90, $\alpha = .10$ and $\alpha/2 = .10/2 = .05$. From Table III, Appendix A, $z_{.05} = 1.645$. The confidence interval is:

$$\bar{x} \pm z_{\alpha/2} \frac{s}{\sqrt{n}} \Rightarrow 113 \pm 1.645 \frac{21}{\sqrt{84}} \Rightarrow 113 \pm 3.77 \Rightarrow (109.23, 116.77)$$

6.13 a. Using the fact that the range of a data set is approximately 4s, then

$$s \approx \frac{\text{range}}{4} = \frac{5.6 - 3}{4} = .65$$

b. For confidence coefficient .90, $\alpha = 1 - .90 = .10$ and $\alpha/2 = .10/2 = .05$. From Table III, Appendix A, $z_{.05} = 1.645$. The 90% confidence interval for μ is:

$$\bar{x} \pm z_{.05}\sigma_{\bar{x}}$$

$$\Rightarrow \bar{x} \pm 1.645 \frac{\sigma}{\sqrt{n}}$$

$$\Rightarrow 4.3 \pm 1.645 \frac{.65}{\sqrt{48}}$$

$$\Rightarrow 4.3 \pm .154 \Rightarrow (4.146, 4.454)$$

6.15 For confidence coefficient .95, α = .05 and $\alpha/2$ = .025. From Table III, Appendix A, $z_{.025}$ = 1.96. The confidence interval is:

$$\bar{x} \pm z_{\alpha/2}\frac{s}{\sqrt{n}} \Rightarrow 19.8 \pm 1.96\frac{\sqrt{25}}{\sqrt{64}} \Rightarrow 19.8 \pm 1.225 \Rightarrow (18.575, 21.025)$$

We are 95% confident the mean number of admissions per 24-hour period is between 18.575 and 21.025.

6.17 a. Note: $E(x) = \mu$.

For confidence coefficient .95, α = .05 and $\alpha/2$ = .025. From Table III, Appendix A, $z_{.025}$ = 1.96. The confidence interval is:

$$\bar{x} \pm z_{\alpha/2}\frac{s}{\sqrt{n}} \Rightarrow 23.43 \pm 1.96\,\frac{10.82}{\sqrt{96}} \Rightarrow 23.43 \pm 2.164$$

$$\Rightarrow (21.266, 25.594)$$

b. We are 95% confident the mean years of service of the top executives in the U.S. banking industry is between 21.266 and 25.594 years.

c. We must assume (using the Central Limit Theorem) that the population of the $\bar{x}$'s is approximately normal.

6.19 For confidence coefficient .95, α = .05 and $\alpha/2$ = .025. From Table III, Appendix A, $z_{.025}$ = 1.96.

$$\text{The sample size is } n = \frac{4(z_{\alpha/2})^2\sigma^2}{W^2} = \frac{4(1.96)^2 6.1}{.2^2} = 2343.4 \approx 2344$$

This is much larger than n = 586 for Exercise 6.18. The difference is that in Exercise 6.18, the bound or half-interval width is .2, while in this problem the whole-interval width is .2.

6.21 a. Range = 27 - 24 = 3. $\sigma \approx \frac{\text{Range}}{4} = \frac{3}{4} = .75$

For confidence coefficient .90, α = .10 and $\alpha/2$ = .05. From Table III, Appendix A, $z_{.05}$ = 1.645.

$$\text{The sample size is } n = \frac{z^2_{\alpha/2}\sigma^2}{B^2} = \frac{1.645^2(.75^2)}{.15^2} = 67.7 \approx 68$$

b. $\sigma \approx \frac{\text{Range}}{6} = \frac{3}{6} = .5$

$$\text{The sample size is } n = \frac{z^2_{\alpha/2}\sigma^2}{B^2} = \frac{1.645^2(.5^2)}{.15^2} = 30.07 \approx 31$$

6.23 a. For confidence coefficient .90, α = .10 and $\alpha/2$ = .05. From Table III, Appendix A, $z_{.05}$ = 1.645.

The sample size is $n = \frac{z^2_{\alpha/2}\sigma^2}{B^2} = \frac{1.645^2(2^2)}{.1^2} = 1082.4 \approx 1083$

b. In part (a), we found n = 1083. If we used an n of only 100, the width of the confidence interval for μ would be wider since we would be dividing by a smaller number.

c. We know $B = \frac{z_{\alpha/2}\sigma}{\sqrt{n}} \Rightarrow z_{\alpha/2} = \frac{B\sqrt{n}}{\sigma} = \frac{.1\sqrt{100}}{2} = .5$

$P(-.5 \leq z \leq .5) = .1915 + .1915 = .3830$. Thus, the level of confidence in approximately 38.3%.

6.25 For confidence coefficient .90, α = .10 and $\alpha/2$ = .05. From Table III, Appendix A, $z_{.05}$ = 1.645.

The sample size is $n = \frac{z^2_{\alpha/2}\sigma^2}{B^2} = \frac{1.645^2(21^2)}{1.5^2} = 530.4 \approx 531$

6.27 $\sigma \approx \frac{\text{Range}}{4} = \frac{180 - 60}{4} = 30$

For confidence coefficient .95, α = .05 and $\alpha/2$ = .05/2 =.025. From Table III, Appendix A, $z_{.025}$ = 1.96.

The sample size is $n = \frac{z^2_{\alpha/2}\sigma^2}{B^2} = \frac{1.96^2(30^2)}{5^2} = 138.3 \approx 139$

6.29 a. If x is normally distributed, the sampling distribution of $\bar{x}$ is normal, regardless of the sample size.

b. If nothing is known about the distribution of x, the sampling distribution of $\bar{x}$ is approximately normal if n is sufficiently large. If n is not large, the distribution of $\bar{x}$ is unknown if the distribution of x is not known.

6.31 First, we must compute $\bar{x}$ and s.

$$\bar{x} = \frac{\sum x}{n} = \frac{29}{6} = 4.833$$

$$s^2 = \frac{\sum x^2 - \frac{(\sum x)^2}{n}}{n - 1} = \frac{163 - \frac{(29)^2}{6}}{6 - 1} = \frac{22.8333}{5} = 4.5667$$

$$s = \sqrt{4.5667} = 2.1370$$

a. For confidence coefficient .90, $\alpha = 1 - .90 = .10$ and $\alpha/2 = .10/2 = .05$. From Table IV, Appendix A, with $df = n - 1 = 6 - 1 = 5$, $t_{.05} = 2.015$. The 90% confidence interval is:

$$\bar{x} \pm t_{.05} \frac{s}{\sqrt{n}}$$

$$\Rightarrow 4.833 \pm 2.015 \frac{2.137}{\sqrt{6}}$$

$$\Rightarrow 4.833 \pm 1.758 \Rightarrow (3.075,\ 6.591)$$

b. For confidence coefficient .95, $\alpha = 1 - .95 = .05$ and $\alpha/2 = .05/2 = .025$. From Table IV, Appendix A, with $df = n - 1 = 6 - 1 = 5$, $t_{.025} = 2.571$. The 95% confidence interval is:

$$\bar{x} \pm t_{.025} \frac{s}{\sqrt{n}}$$

$$\Rightarrow 4.833 \pm 2.571 \frac{2.137}{\sqrt{6}}$$

$$\Rightarrow 4.833 \pm 2.243 \Rightarrow (2.590,\ 7.076)$$

c. For confidence coefficient .99, $\alpha = 1 - .99 = .01$ and $\alpha/2 = .01/2 = .005$. From Table IV, Appendix A, with $df = n - 1 = 6 - 1 = 5$, $t_{.005} = 4.032$. The 99% confidence interval is:

$$\bar{x} \pm t_{.005} \frac{s}{\sqrt{n}}$$

$$\Rightarrow 4.833 \pm 4.032 \frac{2.137}{\sqrt{6}}$$

$$\Rightarrow 4.833 \pm 3.518 \Rightarrow (1.315,\ 8.351)$$

d. a) For confidence coefficient .90, $\alpha = 1 - .90 = .10$ and $\alpha/2 = .10/2 = .05$. From Table IV, Appendix A, with $df = n - 1 = 25 - 1 = 24$, $t_{.05} = 1.711$. The 90% confidence interval is:

$$\bar{x} \pm t_{.05} \frac{s}{\sqrt{n}}$$

$$\Rightarrow 4.833 \pm 1.711 \frac{2.137}{\sqrt{25}}$$

$$\Rightarrow 4.833 \pm .731 \Rightarrow (4.102,\ 5.564)$$

b) For confidence coefficient .95, $\alpha = 1 - .95 = .05$ and $\alpha/2 = .05/2 = .025$. From Table IV, Appendix A, with $df = n - 1 = 25 - 1 = 24$, $t_{.025} = 2.064$. The 95% confidence interval is:

$$\bar{x} \pm t_{.025}\frac{s}{\sqrt{n}}$$

$$\Rightarrow 4.833 \pm 2.064\,\frac{2.137}{\sqrt{25}}$$

$$\Rightarrow 4.833 \pm .882 \Rightarrow (3.951,\ 5.715)$$

c) For confidence coefficient .99, $\alpha = 1 - .99 = .01$ and $\alpha/2 = .01/2 = .005$. From Table IV, Appendix A, with $df = n - 1 = 25 - 1 = 24$, $t_{.005} = 2.797$. The 99% confidence interval is:

$$\bar{x} \pm t_{.005}\frac{s}{\sqrt{n}}$$

$$\Rightarrow 4.833 \pm 2.797\,\frac{2.137}{\sqrt{25}}$$

$$\Rightarrow 4.833 \pm 1.195 \Rightarrow (3.638,\ 6.028)$$

Increasing the sample size decreases the width of the confidence interval.

6.33 a. For confidence coefficient .95, $\alpha = .05$ and $\alpha/2 = .025$. From Table IV, Appendix A, with $df = n - 1 = 21 - 1 = 20$, $t_{.025} = 2.086$. The confidence interval is:

$$\bar{x} \pm t_{.025}\frac{s}{\sqrt{n}} \Rightarrow 52.6 \pm 2.086\left(\frac{3.22}{\sqrt{21}}\right) \Rightarrow 52.6 \pm 1.47 \Rightarrow (51.13,\ 54.07)$$

b. We are 95% confident the mean pulse rate of all American adult males who jog at least 15 miles per week is between 51.13 and 54.07 beats per minute.

c. We must assume the population of pulse rates of all American adult males who jog at least 15 miles per week is normal.

6.35 Some preliminary calculations are:

$$\bar{x} = \frac{\sum x}{n} = \frac{269}{7} = 38.43 \qquad s^2 = \frac{\sum x^2 - \frac{(\sum x)^2}{n}}{n-1} = \frac{14049 - \frac{269^2}{7}}{7-1} = 618.619$$

$$s = \sqrt{618.619} = 24.8721$$

a. For confidence coefficient .90, α = .05 and $\alpha/2$ = .05. From Table IV, Appendix A, with df = n - 1 = 7 - 1 = 6, $t_{.05}$ = 1.943. The confidence interval is:

$$\bar{x} \pm t_{.05} \frac{s}{\sqrt{n}} \Rightarrow 38.43 \pm 1.943\left(\frac{24.8721}{\sqrt{7}}\right) \Rightarrow 38.43 \pm 18.27$$
$$\Rightarrow (20.16,\ 56.70)$$

b. The distribution of times until the blood clot dissolves may not be normal because the time may depend on the size of the clot and the actual individual.

c. If the distribution of x is not normal, the level of confidence for the confidence interval will not be 90%, but probably something smaller.

6.37 Some preliminary calculations are:

$$\bar{x} = \frac{\sum x}{n} = \frac{811{,}783}{12} = 67{,}648.583$$

$$s^2 = \frac{\sum x^2 - \frac{(\sum x)^2}{n}}{n - 1} = \frac{58{,}958{,}888{,}430 - \frac{811783^2}{12}}{12 - 1} = 367{,}538{,}046.4$$

$$s = \sqrt{367538046.4} = 19{,}171.2818$$

a. The population is the set of all outstanding principal balances of all home mortgages foreclosed by the bank due to default by the borrower during the last three years. This distribution must be normal.

b. For confidence coefficient .90, α = .10 and $\alpha/2$ = .05. From Table IV, Appendix A, with df = n - 1 = 12 - 1 = 11, $t_{.05}$ = 1.796. The confidence interval is:

$$\bar{x} \pm t_{.05} \frac{s}{\sqrt{n}} \Rightarrow 67{,}648.583 \pm 1.796\left(\frac{19{,}171.2818}{\sqrt{12}}\right)$$
$$\Rightarrow 67{,}648.583 \pm 9939.553 \Rightarrow (57{,}709.030,\ 77{,}588.136)$$

c. We are 90% confident the mean outstanding principal balance of all home mortgages foreclosed by the bank due to default is between \$57,709.03 and \$77,588.136.

6.39 a. For confidence coefficient .95, $\alpha = 1 - .95 = .05$ and $\alpha/2 = .05/2 = .025$. From Table IV, Appendix A, with $df = n - 1 = 23 - 1 = 22$, $t_{.025} = 2.074$. The 95% confidence interval is:

$$\bar{x} \pm t_{.025} \frac{s}{\sqrt{n}}$$

$$\Rightarrow 135 \pm 2.074 \frac{32}{\sqrt{23}}$$

$$\Rightarrow 135 \pm 13.839 \Rightarrow (121.161, 148.839)$$

b. We must assume a random sample was selected and that the population of all health insurance costs per worker per month is normally distributed.

c. "95%" confidence interval" means that if repeated samples of size 23 were selected from the population and 95% confidence intervals formed, 95% of all confidence intervals will contain the true value of μ.

6.41 The sampling distribution of $\hat{p}$ is approximately normal with mean $\mu = p$ and standard deviation $\sigma_{\hat{p}} = \sqrt{\frac{pq}{n}}$

6.43 The sample size is sufficiently large if $\hat{p} \pm 3\sigma_{\hat{p}}$ does not include 0 or 1.

a. $\hat{p} \pm 3\sigma_{\hat{p}} \approx .05 \pm 3\sqrt{\frac{.05(.95)}{500}} \Rightarrow .05 \pm .029 \Rightarrow (.021, .079)$

Since this interval does not contain 0 or 1, the sample size is sufficiently large.

b. $\hat{p} \pm 3\sigma_{\hat{p}} \approx .05 \pm 3\sqrt{\frac{.05(.95)}{100}} \Rightarrow .05 \pm .065 \Rightarrow (-.015, .115)$

Since the interval contains 0, the sample size is not sufficiently large.

c. $\hat{p} \pm 3\sigma_{\hat{p}} \approx .5 \pm 3\sqrt{\frac{.5(.5)}{10}} \Rightarrow .5 \pm .474 \Rightarrow (.026, .974)$

Since this interval does not contain 0 or 1, the sample size is sufficiently large.

d. $\hat{p} \pm 3\sigma_{\hat{p}} \approx .3 \pm 3\sqrt{\frac{.3(.7)}{10}} \Rightarrow .3 \pm .435 \Rightarrow (-.135, .735)$

Since the interval contains 0, the sample size is not sufficiently large.

6.45 $\hat{p} = \frac{x}{n} = \frac{76}{122} = .623$

a. We first check to see if the sample size is sufficiently large.

$$\hat{p} \pm 3\sigma_{\hat{p}} \approx \hat{p} \pm 3\sqrt{\frac{(\hat{p}\hat{q})}{n}}$$

$$\Rightarrow .623 \pm 3\sqrt{\frac{(.623)(.377)}{122}}$$

$$\Rightarrow .623 \pm .132 \Rightarrow (.491, .755)$$

Since the interval is wholly contained in the interval (0, 1), we may conclude that the normal approximation is reasonable.

For confidence coefficient .95, $\alpha = 1 - .95 = .05$ and $\alpha/2 = .05/2 = .025$. From Table III, Appendix A, $z_{.025} = 1.96$. The 95% confidence interval is:

$$\hat{p} \pm z_{.025}\sqrt{\frac{(\hat{p}\hat{q})}{n}}$$

$$\Rightarrow .623 \pm 1.96\sqrt{\frac{(.623)(.377)}{122}}$$

$$\Rightarrow .623 \pm .086 \Rightarrow (.537, .709)$$

b. We are 95% confident the proportion of all Illinois law firms who used microcomputers at the time of the survey is between .537 and .709.

c. "95% confidence interval" means that if repeated samples of size 122 were selected from the population and 95% confidence intervals formed, 95% of all confidence intervals will contain the true value of p.

d. Probably not. The sample was selected only from Illinois law firms. The only way the interval could be used to estimate the proportion of all U.S. law firms who were using microcomputers at the time of the survey would be if the Illinois law firms are very similar to all U.S. law firms.

6.47 a. The population of interest is the collection of responses to the question "Are you under 45 years of age?" obtained from all owners and managers of small U.S. businesses.

b. The sample size is large enough if $\hat{p} \pm 3\sigma_{\hat{p}}$ lies within the interval (0, 1).

$$\hat{p} \pm 3\sigma_{\hat{p}} \Rightarrow \hat{p} \pm 3\sqrt{\frac{pq}{n}} \approx \hat{p} \pm 3\sqrt{\frac{\hat{p}\hat{q}}{n}}$$

$$\Rightarrow .4 \pm 3\sqrt{\frac{.4(.6)}{258}}$$

$$\Rightarrow .4 \pm .091 = (.309, .491)$$

Since the interval lies completely in the interval (0, 1), the normal approximation will be adequate.

c. For confidence coefficient .95, $\alpha = .05$ and $\alpha/2 = .025$. From Table III, Appendix A, $z_{.025} = 1.96$. The 95% confidence interval is:

$$\hat{p} \pm z_{.025}\sqrt{\frac{pq}{n}}$$

$$\Rightarrow \hat{p} \pm 1.96\sqrt{\frac{\hat{p}\hat{q}}{n}}$$

$$\Rightarrow .4 \pm 1.96\sqrt{\frac{.4(.6)}{258}}$$

$$\Rightarrow .4 \pm .06 \Rightarrow (.34, .46)$$

d. The interval would be narrower for $\hat{p} = .30$ rather than $\hat{p} = .40$. As $\hat{p}$ approaches .5, the standard error, $\sigma_{\hat{p}}$, gets larger; therefore, a smaller $\hat{p}$ produces a smaller $\sigma_{\hat{p}}$ and a narrower confidence interval.

6.49 a. For KOCO, $\hat{p} = \frac{92}{119} = .773$

For WCBS, $\hat{p} = \frac{21}{35} = .6$

For WXIA, $\hat{p} = \frac{79}{177} = .446$

The sample size is large enough if $\hat{p} \pm 3\sigma_{\hat{p}}$ lies within the interval (0, 1).

For KOCO, $\hat{p} \pm 3\sigma_{\hat{p}} \Rightarrow \hat{p} \pm 3\sqrt{\frac{pq}{n}} \approx \hat{p} \pm 3\sqrt{\frac{\hat{p}\hat{q}}{n}}$

$$\Rightarrow .773 \pm 3\sqrt{\frac{.773(.227)}{119}}$$

$$\Rightarrow .773 \pm .1152 \Rightarrow (.6578, .8882)$$

Since the interval lies completely in the interval (0, 1), the normal approximation will be adequate.

For WCBS, $\hat{p} \pm 3\sigma_{\hat{p}} \Rightarrow \hat{p} \pm 3\sqrt{\frac{pq}{n}} \approx \hat{p} \pm 3\sqrt{\frac{\hat{p}\hat{q}}{n}}$

$$\Rightarrow .6 \pm 3\sqrt{\frac{.6(.4)}{35}}$$

$$\Rightarrow .6 \pm .2484 \Rightarrow (.3516, .8484)$$

Since the interval lies completely in the interval (0, 1), the normal approximation will be adequate.

For WXIA, $\hat{p} \pm 3\sigma_{\hat{p}} \Rightarrow \hat{p} \pm 3\sqrt{\frac{pq}{n}} \approx \hat{p} \pm 3\sqrt{\frac{\hat{p}\hat{q}}{n}}$

$$\Rightarrow .446 \pm 3\sqrt{\frac{.446(.554)}{177}}$$

$$\Rightarrow .446 \pm .1121 \Rightarrow (.3339, .5581)$$

Since the interval lies completely in the interval (0, 1), the normal approximation will be adequate.

b. For confidence coefficient .95, $\alpha = .05$ and $\alpha/2 = .025$. From Table III, Appendix A, $z_{.025} = 1.96$. The form of the confidence interval is:

$$\hat{p} \pm z_{.025}\sqrt{\frac{\hat{p}\hat{q}}{n}}$$

For KOCO, $.773 \pm 1.96\sqrt{\frac{.773(.227)}{119}} \Rightarrow .773 \pm .075 \Rightarrow (.698, .848)$

For WCBS, $.6 \pm 1.96\sqrt{\frac{.6(.4)}{35}} \Rightarrow .6 \pm .162 \Rightarrow (.438, .762)$

For WXIA, $.446 \pm 1.96\sqrt{\frac{.446(.554)}{177}} \Rightarrow .446 \pm .073 \Rightarrow (.373, .519)$

c. $\hat{p} = \frac{92 + 21 + 79}{331} = .58$

For confidence coefficient .90, $\alpha = 1 - .90 = .10$ and $\alpha/2 = .10/2 = .05$. From Table III, Appendix A, $z_{.05} = 1.645$. The 90% confidence interval is:

$$\hat{p} \pm z_{.05}\sqrt{\frac{\hat{p}\hat{q}}{n}}$$

$$\Rightarrow .58 \pm 1.645\sqrt{\frac{.58(.42)}{331}}$$

$$\Rightarrow .58 \pm .045 \Rightarrow (.535, .625)$$

6.51 The "margin of error" is approximately $\pm 1.96\sqrt{\frac{\hat{p}\hat{q}}{n}}$ with 95% confidence.

For the 46%, $\Rightarrow \pm 1.96\sqrt{\frac{.46(.54)}{505}} \Rightarrow \pm .043$

For the 37%, $\Rightarrow \pm 1.96\sqrt{\frac{.37(.63)}{505}} \Rightarrow \pm .042$

For the 30%, $\Rightarrow \pm 1.96\sqrt{\frac{.3(.7)}{505}} \Rightarrow \pm .040$

6.53 a. For probability .90, $\alpha = .10$ and $\alpha/2 = .05$.

From Table III, Appendix A, $z_{.05} = 1.645$.

$$n = \frac{(z_{\alpha/2})^2 pq}{B^2} = \frac{1.645^2(.8)(.2)}{.02^2} = 1082.4 \approx 1083$$

b. If we do not know p, use p = .5. It will give the most conservative estimate.

$$n = \frac{(z_{\alpha/2})^2 pq}{B^2} = \frac{1.645^2(.5)(.5)}{.02^2} = 1691.3 \approx 1692$$

6.55 For confidence coefficient .90, $\alpha = .10$ and $\alpha/2 = .05$. From Table III, Appendix A, $z_{.05} = 1.645$.

We know $\hat{p}$ is in the middle of the interval, so $\hat{p} = \frac{.54 + .26}{2} = .4$

The confidence interval is

$$\hat{p} \pm z_{.05}\sqrt{\frac{\hat{p}\hat{q}}{n}} \Rightarrow .4 \pm 1.645\sqrt{\frac{.4(.6)}{n}}$$

We know $.4 - 1.645\sqrt{\frac{.4(.6)}{n}} = .26$

$$\Rightarrow .4 - \frac{.8059}{\sqrt{n}} = .26$$

$$\Rightarrow .4 - .26 = \frac{.8059}{\sqrt{n}} \Rightarrow \sqrt{n} = \frac{.8059}{.14} = 5.756$$

$$\Rightarrow n = 5.756^2 = 33.1 \approx 34$$

6.57 For confidence coefficient .95, α = .05 and $\alpha/2$ = .05/2 = .025. From Table III, Appendix A, $z_{.025}$ = 1.96. For this study,

$$\hat{p} = \frac{190}{532} = .36.$$ Using $\hat{p}$ to estimate p, we get

$$n = \frac{(z_{\alpha/2})^2 pq}{B^2} \approx \frac{1.96^2(.36)(.64)}{.03^2} = 983.4 \approx 984$$

The sample size needed to be 984 instead of 532.

6.59 For confidence coefficient .95, α = .05 and $\alpha/2$ = .025. From Table III, Appendix A, $z_{.025}$ = 1.96.

$$n = \frac{4(z_{\alpha/2})^2 pq}{W^2} = \frac{4(1.96)^2(.3)(.7)}{.01^2} = 32{,}269.4 \approx 32{,}270$$

6.61 With probability .90, α = .10 and $\alpha/2$ = .05. From Table III, Appendix A, $z_{.05}$ = 1.645.

$$n = \frac{(z_{\alpha/2})^2 pq}{B^2} \approx \frac{1.645^2(.05)(.95)}{.02^2} = 321.3 \approx 322$$

6.63 a. For a small sample from a normal distribution with unknown standard deviation, we use the t statistic. For confidence coefficient .95, α = 1 - .95 = .05 and $\alpha/2$ = .05/2 = .025. From Table IV, Appendix A, with df = n - 1 = 23 - 1 = 22, $t_{.025}$ = 2.074.

b. For a large sample from a distribution with an unknown standard deviation, we can estimate the population standard deviation with s and use the z statistic. For confidence coefficient .95, α = 1 - .95 = .05 and $\alpha/2$ = .05/2 = .025. From Table III, Appendix A, $z_{.025}$ = 1.96.

c. For a small sample from a normal distribution with known standard deviation, we use the z statistic. For confidence coefficient .95, α = 1 - .95 = .05 and $\alpha/2$ = .05/2 = .025. From Table III, Appendix A, $z_{.025}$ = 1.96.

d. For a large sample from a distribution about which nothing is known, we can estimate the population standard deviation with s and use the z statistic. For confidence coefficient .95, α = 1 - .95 = .05 and $\alpha/2$ = .05/2 = .025. From Table III, Appendix A, $z_{.025}$ = 1.96.

e. For a small sample from a distribution about which nothing is known, we can use neither z nor t.

6.65 a. First, check to see if the normal approximation will be adequate.

$$p \pm 3\sigma_{\hat{p}} \Rightarrow p \pm 3\sqrt{\frac{pq}{n}} \Rightarrow \hat{p} \pm 3\sqrt{\frac{\hat{p}\hat{q}}{n}} \text{ where } \hat{p} = \frac{38}{60} = .633$$

$$\Rightarrow .633 \pm 3\sqrt{\frac{.633(.367)}{60}}$$

$$\Rightarrow .633 \pm .187 \Rightarrow (.446, .820)$$

Since the interval lies completely in the interval (0, 1), the normal approximation will be adequate.

For confidence coefficient .90, $\alpha = .10$ and $\alpha/2 = .05$. From Table III, Appendix A, $z_{.05} = 1.645$. The 90% confidence interval is:

$$\hat{p} \pm z_{.05}\sqrt{\frac{pq}{n}}$$

$$\Rightarrow \hat{p} \pm 1.645\sqrt{\frac{\hat{p}\hat{q}}{n}}$$

$$\Rightarrow .633 \pm 1.645\sqrt{\frac{.633(.367)}{60}}$$

$$\Rightarrow .633 \pm .102 \Rightarrow (.531, .735)$$

b. We are 90% confident the true proportion of employees who prefer an on-site program lies in the interval .531 to .735.

c. If repeated samples of size 60 were selected from the population and 90% confidence intervals formed, 90% of all intervals formed will contain the true value of p.

6.67 a. For confidence coefficient .95, $\alpha = .05$ and $\alpha/2 = .025$. From Table III, Appendix A, $z_{.025} = 1.96$. The confidence interval is:

$$\hat{p} \pm z_{\alpha/2}\sqrt{\hat{p}\hat{q}/n} \Rightarrow .075 \pm 1.96\sqrt{(.075)(.925)/200}$$

$$\Rightarrow .075 \pm .037 \Rightarrow (.038, .112)$$

The bank can be 95% confident that the actual proportion of the bank's savings accounts on whose balance the bank and customer disagree is between .038 and .112.

b. Since the confidence interval in part (a) contains some values which are larger than .10, the interval does not support the claim that the true fraction of accounts on which there is disagreement is no larger than .10. The true fraction could be as large as .112 from the interval in part (a).

6.69 a. For confidence coefficient .99, $\alpha = .01$ and $\alpha/2 = .005$. From Table IV, Appendix A, with df $= n - 1 = 20 - 1 = 19$, $t_{.005} = 2.861$. The confidence interval is:

$$\bar{x} \pm t_{.005}\left(\frac{s}{\sqrt{n}}\right) \Rightarrow 54 \pm 2.861\left(\frac{9.59}{\sqrt{20}}\right) \Rightarrow 54 \pm 6.135$$

$$\Rightarrow (47.865, 60.135)$$

b. A measure that might be of more interest than the mean is the maximum depth. One would need to buy equipment capable of handling any depth received. Thus, the maximum rather than the mean may be more important.

6.71 $\hat{p} = \frac{10}{80} = .125$

For confidence coefficient .99, $\alpha = .01$ and $\alpha/2 = .005$. From Table III, Appendix A, $z_{.005} = 2.58$. The confidence interval is:

$$\hat{p} \pm z_{\alpha/2}\sqrt{\hat{p}\hat{q}/n} \Rightarrow .125 \pm 2.58\sqrt{(.125)(.875)/80}$$

$$\Rightarrow .125 \pm .095 \Rightarrow (.030, .220)$$

6.73 For confidence coefficient .95, $\alpha = .05$ and $\alpha/2 = .025$. From Table IV, Appendix A, with df $= n - 1 = 9 - 1 = 8$, $t_{.025} = 2.306$. The confidence interval is:

$$\bar{x} \pm t_{.025}\frac{s}{\sqrt{n}} \Rightarrow 26.4 \pm 2.306\left(\frac{2}{\sqrt{9}}\right) \Rightarrow 26.4 \pm 1.54 \Rightarrow (24.86, 27.94)$$

We are 95% confident the mean nicotine content per cigarette for this brand is between 24.86 mg and 27.94 mg.

6.75 $\hat{p} = \frac{x}{n} = \frac{212}{346} = .613$

For confidence coefficient .95, $\alpha = .05$ and $\alpha/2 = .05/2 = .025$. From Table III, Appendix A, $z_{.025} = 1.96$. The 95% confidence interval is:

$$\hat{p} \pm z_{.025}\sqrt{\frac{\hat{p}\hat{q}}{n}}$$

$$\Rightarrow .613 \pm 1.96\sqrt{\frac{(.613)(.387)}{346}}$$

$$\Rightarrow .613 \pm .051 \Rightarrow (.562, .664)$$

We are 95% confident the proportion of households in the city who are using available recycling facilities is between .562 and .664.

6.77 To compute the needed sample size, use:

$$n = \frac{4(z_{\alpha/2})^2 pq}{W^2} \quad \text{where } z_{.025} = 1.96 \text{ from Table III, Appendix A.}$$

$$\text{Thus, } n = \frac{4(1.96)^2(.094)(.906)}{.04^2} \approx 817.9 \approx 818$$

6.79 a. To compute the necessary sample size, use:

$$n = \frac{(z_{\alpha/2})^2 \sigma^2}{B^2} \quad \text{where } \alpha = 1 - .99 = .01 \text{ and } \alpha/2 = .01/2 = .005$$

From Table III, Appendix A, $z_{.005} = 2.575$. Thus,

$$n = \frac{(2.575)^2 11.34^2}{1^2} = 852.7 \approx 853$$

b. We would have to assume the sample was a random sample.

CHAPTER 7

INFERENCES BASED ON A SINGLE SAMPLE: TESTS OF HYPOTHESES

7.1 The null hypothesis is the "status quo" hypothesis, while the alternative hypothesis is the research hypothesis.

7.3 The "level of significance" of a test is α. This is the probability that the test statistic will fall in the rejection region when the null hypothesis is true.

7.5 The four possible results are:

1. Rejecting the null hypothesis when it is true. This would be a Type I error.

2. Accepting the null hypothesis when it is true. This would be a correct decision.

3. Rejecting the null hypothesis when it is false. This would be a correct decision.

4. Accepting the null hypothesis when it is false. This would be a Type II error.

7.7 When you reject the null hypothesis in favor of the alternative hypothesis, this does not prove the alternative hypothesis is correct. We are $100(1 - \alpha)\%$ confident that there is sufficient evidence to conclude that the alternative hypothesis is correct.

If we were to repeatedly draw samples from the population and perform the test each time, approximately $100(1 - \alpha)\%$ of the tests performed would yield the correct decision.

7.9 a. Since the company must give proof the drug is safe, the null hypothesis would be the drug is unsafe. The alternative hypothesis would be the drug is safe.

b. A Type I error would be concluding the drug is safe when it is not safe. A Type II error would be concluding the drug is not safe when it is. α is the probability of concluding the drug is safe when it is not. β is the probability of concluding the drug is not safe when it is.

c. In this problem, it would be more important for α to be small. We would want the probability of concluding the drug is safe when it is not to be as small as possible.

7.11 a. $z = \frac{\bar{x} - \mu}{\sigma_{\bar{x}}} = \frac{212 - 200}{80/\sqrt{100}} = 1.50$

The decision rule is: Reject H_0 if $z > 1.50$.

b. $\alpha = P(z > 1.50) = .5 - .4332 = .0668$

7.13 a. $\bar{x} = \frac{\sum x}{n} = \frac{20.7}{49} = .4224$ $\quad s^2 = \frac{\sum(x_i - \bar{x})^2}{n - 1} = \frac{2.155}{49 - 1} = .0449$

$s = \sqrt{.0449} = .2119$

H_0: $\mu = .47$
H_a: $\mu < .47$

The test statistic is $z = \frac{\bar{x} - \mu_0}{\sigma_{\bar{x}}} = \frac{.4224 - .47}{.2119/\sqrt{49}} = -1.57$

The rejection region requires $\alpha = .10$ in the lower tail of the z distribution. From Table III, Appendix A, $z_{.10} = 1.28$. The rejection region is $z < -1.28$.

Since the observed value of the test statistic falls in the rejection region ($z = -1.57 < -1.28$), H_0 is rejected. There is sufficient evidence to indicate the mean is less than .47 at $\alpha = .10$.

b. H_0: $\mu = .47$
H_a: $\mu \neq .47$

The test statistic is $z = -1.57$ (see part (b)).

The rejection region requires $\alpha/2 = .10/2 = .05$ in the each tail of the z distribution. From Table III, Appendix A, $z_{.05} = 1.645$. The rejection region is $z < -1.645$ or $z > 1.645$.

Since the observed value of the test statistic does not fall in the rejection region ($z = -1.57 \nless -1.645$), H_0 is not rejected. There is insufficient evidence to indicate the mean is different from .47 at $\alpha = .10$.

7.15 a. The null hypothesis is H_0: $\mu = 35$ mpg. The alternative hypothesis is H_a: $\mu > 35$ mpg.

b. We want to test:

H_0: $\mu = 35$
H_a: $\mu > 35$

The test statistic is $z = \frac{\bar{x} - \mu_0}{\sigma_{\bar{x}}} = \frac{\bar{x} - \mu_0}{\sigma/\sqrt{n}} \approx \frac{\bar{x} - \mu_0}{s/\sqrt{n}} = \frac{36.8 - 35}{6/\sqrt{36}}$

$= 1.8$

The rejection region requires $\alpha = .10$ in the upper tail of the z distribution. From Table III, Appendix A, $z_{.10} = 1.28$. The rejection region is $z > 1.28$.

Since the observed value of the test statistic falls in the rejection region ($z = 1.8 > 1.28$), H_0 is rejected. There is sufficient evidence to support the auto manufacturer's claim that the mean miles per gallon for the car exceeds the EPA estimate of 35 at $\alpha = .10$.

The conclusion is the same as that in Exercise 7.14.

7.17 Some preliminary calculations are:

$$\bar{x} = \frac{\sum x}{n} = \frac{10.099}{40} = .252475$$

$$s^2 = \frac{\sum x^2 - \frac{(\sum x)^2}{n}}{n - 1} = \frac{2.549934 - \frac{10.099^2}{40}}{40 - 1} = .000004973$$

$$s = \sqrt{.000004973} = .00223$$

a. To determine if the process is not operating satisfactorily, we test:

H_0: $\mu = .250$
H_a: $\mu \neq .250$

The test statistic is $z = \frac{\bar{x} - \mu_0}{\sigma_{\bar{x}}} = \frac{\bar{x} - \mu_0}{\sigma/\sqrt{n}} \approx \frac{\bar{x} - \mu_0}{s/\sqrt{n}}$

$$= \frac{.252475 - .250}{.00223/\sqrt{40}} = 7.02$$

The rejection region requires $\alpha/2 = .05/2 = .025$ in each tail of the z distribution. From Table III, Appendix A, $z_{.025} = 1.96$. The rejection region is $z < -1.96$ or $z > 1.96$.

Since the observed value of the test statistic falls in the rejection region ($z = 7.02 > 1.96$), H_0 is rejected. There is sufficient evidence to indicate the process is not operating satisfactorily at $\alpha = .05$.

b. α = Probability of rejecting H_0 when H_0 is true. If H_0 is true ($\mu = .250$), the process is in control. If H_0 is rejected when it is true, the process is declared not operating satisfactorily when in fact it is. This is a risk to the producer.

β = Probability of accepting H_0 when H_0 is false. If H_0 is false ($\mu \neq .250$), the process is not operating satisfactorily. If H_0 is accepted when it is false, the process is declared to be operating satisfactorily when it is not. This is a risk to the consumer.

7.19 H_0: $\mu = 900$
H_a: $\mu > 900$

The test statistic is $z = \dfrac{\bar{x} - \mu_0}{\sigma_{\bar{x}}} = \dfrac{920 - 900}{80/\sqrt{64}} = 2.00$

The rejection region requires $\alpha = .05$ in the upper tail of the z distribution. From Table III, Appendix A, $z_{.05} = 1.645$. The rejection region is $z > 1.645$.

Since the observed value of the test statistic falls in the rejection region ($2.00 > 1.645$), H_0 is rejected. There is sufficient evidence to indicate that the mean life of the new brand of bulbs is in excess of 900 hours at $\alpha = .05$. The University will purchase the new brand of fluorescent bulbs.

7.21 a. Since the p-value = .10 is greater than $\alpha = .05$, H_0 is not rejected.

b. Since the p-value = .05 is less than $\alpha = .10$, H_0 is rejected.

c. Since the p-value = .001 is less than $\alpha = .01$, H_0 is rejected.

d. Since the p-value = .05 is greater than $\alpha = .025$, H_0 is not rejected.

e. Since the p-value = .45 is greater than $\alpha = .10$, H_0 is not rejected.

7.23 The smallest value of α for which the null hypothesis would be rejected is just greater than .06.

7.25 p-value $= P(z \geq 2.26) = .5 - P(0 < z < 2.26)$
$= .5 - .4881$
$= .0119$

7.27 p-value $= P(z \geq 2.08) + P(z \leq -2.08) = (.5 - .4812)2 = .0376$

7.29 a. To test the manufacturer's claim, we test:

H_0: $\mu = 3.5$
H_a: $\mu < 3.5$

The test statistic is $z = \frac{\bar{x} - \mu_0}{\sigma_{\bar{x}}} = \frac{3.3 - 3.5}{\frac{66/60}{\sqrt{50}}} = -1.29$

The rejection region requires $\alpha = .05$ in the lower tail of the z distribution. From Table III, Appendix A, $z_{.05} = 1.645$. The rejection region is $z < -1.645$.

Since the observed value of the test statistic does not fall in the rejection region ($z = -1.29 \not< -1.645$), H_0 is not rejected. There is insufficient evidence to support the manufacturer's claim at $\alpha = .05$.

b. p-value $= P(z \leq -1.29) = .5 - .4015 = .0985$

c. The smaller the p-value, the more support for the manufacturer's claim. The p-value measures the probability of observing your test statistic or anything more unusual if H_0 is true. As the p-value gets smaller, there is more evidence to reject H_0.

7.31 From Exercise 7.14, the test statistic is:

$$z = \frac{\bar{x} - \mu_0}{\sigma_{\bar{x}}} = \frac{36.8 - 35}{6/\sqrt{36}} = 1.8$$

p-value $= P(z \geq 1.8) = .5 - .4641 = .0359$

The probability of observing a test statistic of $z = 1.8$ or larger is .0359. This is a rare event if H_0 is true.

7.33 From Exercise 7.18, the test statistic is $z = .87$.

p-value $= P(z \geq .87) = .5 - .3078 = .1922$

7.35 From Exercise 7.19, the test statistic is

$$z = \frac{\bar{x} - \mu_0}{\sigma_{\bar{x}}} = \frac{920 - 900}{80/\sqrt{64}} = 2$$

p-value $= P(z \geq 2) = .5 - .4772 = .0228$

7.37 We should use the t-distribution in testing a hypothesis about a population mean if the sample size is small and if the population being sampled from is normal.

7.39 Some preliminary calculations are:

$$\bar{x} = \frac{\sum x}{n} = \frac{24}{5} = 4.8 \qquad s^2 = \frac{\sum x^2 - \frac{(\sum x)^2}{n}}{n - 1} = \frac{126 - \frac{24^2}{5}}{5 - 1} = \frac{10.8}{4} = 2.7$$

$s = \sqrt{2.7} = 1.6432$

a. $H_0: \mu = 6$
$H_a: \mu < 6$

The test statistic is $t = \dfrac{\bar{x} - \mu_0}{s/\sqrt{n}} = \dfrac{4.8 - 6}{1.6432/\sqrt{5}} = -1.63$

The necessary assumption is that the population is normal.

The rejection region requires $\alpha = .05$ in the lower tail of the t distribution with df = n - 1 = 5 - 1 = 4. From Table IV, Appendix A, $t_{.05} = 2.132$. The rejection region is $t < -2.132$.

Since the observed value of the test statistic does not fall in the rejection region ($t = -1.63 \nless -2.132$), H_0 is not rejected. There is insufficient evidence to indicate the mean is less than 6 at $\alpha = .05$.

b. $H_0: \mu = 6$
$H_a: \mu \neq 6$

The test statistic is $t = -1.63$ (from (a)).

The assumption is the same as in (a).

The rejection region requires $\alpha/2 = .05/2 = .025$ in each tail of the t distribution with df = n - 1 = 5 - 1 = 4. From Table IV, Appendix A, $t_{.025} = 2.776$. The rejection region is $t < -2.776$ or $t > 2.776$.

Since the observed value of the test statistic does not fall in the rejection region ($t = -1.63 \nless -2.776$), H_0 is not rejected. There is insufficient evidence to indicate the mean is different from 6 at $\alpha = .05$.

c. For part (a), the p-value = $P(t \leq -1.63)$.

From Table IV, with df = 4, $.05 < P(t \leq -1.63) < .10$.

For part (b), the p-value = $P(t \leq -1.63) + P(t \geq 1.63)$.

From Table IV, with df = 4, $2(.05) <$ p-value $< 2(.10)$ or $.10 <$ p-value $< .20$.

7.41 a. The consumer group would test:

$H_0: \mu = 20$
$H_a: \mu < 20$

The test statistic is $t = \frac{\bar{x} - \mu_0}{s/\sqrt{n}} = \frac{19.86 - 20}{.22/\sqrt{10}} = -2.01$

The rejection region requires $\alpha = .05$ in the lower tail of the t distribution with df = n - 1 = 10 - 1 = 9. From Table IV, Appendix A, $t_{.05} = 1.833$. The rejection region is $t < -1.833$.

Since the observed value of the test statistic falls in the rejection region ($-2.01 < -1.833$), H_0 is rejected. There is sufficient evidence to indicate that the mean fill per family-size bottles is less than 20 ounces at $\alpha = .05$.

b. The consumer group, more interested in not being cheated, would be more concerned with a Type II error, i.e., they would be more concerned about the possibility that the bottles are being underfilled, but the reverse decision is made.

c. Since deciding that the bottles are being underfilled when they are not would cause the company to put more catsup in the bottles, the company would be more concerned with a Type I error.

d. For confidence coefficient .90, $\alpha = .10$ and $\alpha/2 = .05$. From Table IV, Appendix A, with df = 9, $t_{.05} = 1.833$. The confidence interval is:

$$\bar{x} \pm t_{\alpha/2}\frac{s}{\sqrt{n}} \Rightarrow 19.86 \pm 1.833\frac{.22}{\sqrt{10}} \Rightarrow 19.86 \pm .13 \Rightarrow (19.73, 19.99)$$

We are 90% confident the mean number of ounces of catsup being dispensed is between 19.73 and 19.99 ounces.

7.43 a. $\bar{x} = \frac{24 + 20 + 22}{3} = 22$

$$s^2 = \frac{(24 - 22)^2 + (20 - 22)^2 + (20 - 20)^2}{3 - 1} = 4$$

$H_0: \mu = 21$
$H_a: \mu > 21$

The test statistic is $t = \frac{\bar{x} - \mu_0}{s/\sqrt{n}} = \frac{22 - 21}{\sqrt{4}/\sqrt{3}} = 0.87$

The rejection region requires $\alpha = .10$ in the upper tail of the t distribution with df = n - 1 = 3 - 1 = 2. From Table IV, Appendix A, $t_{.10} = 1.886$. The rejection region is $z > 1.886$.

Since the observed value of the test statistic does not fall in the rejection region ($0.87 \not> 1.886$), H_0 is not rejected. There is insufficient evidence to indicate that the mean length of the great white sharks off the Bermuda coast is in excess of 21 feet at $\alpha = .10$.

b. Since $0.87 < 1.886 \Rightarrow$ p-value $> .10$.

c. We must be able to assume that the lengths of the great white sharks off the Bermuda coast is a normal random variable and that a random sample has been obtained.

d. Those sharks which are slow/old/near shore/etc. are more likely to be selected, indicating a strong possibility that this is not a random sample.

7.45 a. To evaluate the self-worth of the male heroin addicts, we test:

H_0: $\mu = 48.6$
H_a: $\mu \neq 48.6$

where μ = mean test score

The test statistic is $t = \dfrac{\bar{x} - \mu_0}{s/\sqrt{n}} = \dfrac{44.1 - 48.6}{6.2/\sqrt{25}} = -3.63$

The rejection region requires $\alpha/2 = .01/2 = .005$ in each tail of the t distribution with df $= n - 1 = 25 - 1 = 24$. From Table IV, Appendix A, $t_{.005} = 2.797$. The rejection region is $z < -2.797$ or $z > 2.797$.

Since the observed value of the test statistic falls in the rejection region ($-3.63 < -2.797$), H_0 is rejected. There is sufficient evidence to indicate that the mean self-worth of the male heroin addicts is different from the general male population at $\alpha = .01$.

b. From Table IV, Appendix A, with df = 24,

$$-3.745 < -3.63 < -2.797 \Rightarrow 2(.0005) < \text{p-value} < 2(.001)$$
$$\Rightarrow .001 < \text{p-value} < .002$$

7.47 a. Since the normal distribution is symmetric, the probability that a randomly selected observation exceeds the mean of a normal distribution is .5.

b. By the definition of "median," the probability that a randomly selected observation exceeds the median of a normal distribution is .5.

c. If the distribution is not normal, the probability that a randomly selected observation exceeds the mean depends on the distribution. With the information given, the probability cannot be determined.

d. By definition of "median," the probability that a randomly selected observation exceeds the median of a non-normal distribution is .5.

7.49 a. H_0: $M = 10$
H_a: $M > 10$

The test statistic is S = {Number of observations greater than 10} = 6.

The p-value = $P(x \geq 6)$ where x is a binomial random variable with n = 10 and p = .5. From Table II,

$$\text{p-value} = P(x \geq 6) = 1 - P(x \leq 5) = 1 - .623 = .377$$

Since the p-value = .377 > α = .05, H_0 is not rejected. There is insufficient evidence to indicate the median is greater than 10 at α = .05.

b. H_0: $M = 10$
H_a: $M \neq 10$

S_1 = {Number of observations less than 10} = 4 and
S_2 = {Number of observations greater than 10} = 6

The test statistic is S = larger of S_1 and S_2 = 6.

The p-value = $2P(x \geq 6)$ where x is a binomial random variable with n = 10 and p = .5. From Table II,

$$\text{p-value} = 2P(x \geq 6) = 2(1 - P(x \leq 5) = 2(1 - .623) = .754$$

Since the p-value = .754 > α = .05, H_0 is not rejected. There is insufficient evidence to indicate the median is different than 10 at α = .05.

c. H_0: $M = 18$
H_a: $M < 18$

The test statistic is S = {Number of observations less than 18} = 9

The p-value = $P(x \geq 9)$ where x is a binomial random variable with n = 10 and p = .5. From Table II,

$$\text{p-value} = P(x \geq 9) = 1 - P(x \leq 8 = 1 - .989 = .011$$

Since the p-value = .011 $<$ α = .05, H_0 is rejected. There is sufficient evidence to indicate the median is less than 18 at α = .05.

d. H_0: $M = 18$
H_a: $M \neq 18$

S_1 = {Number of observations less than 18} = 9 and
S_2 = {Number of observations greater than 18} = 1

The test statistic is S = larger of S_1 and S_2 = 9.

The p-value = $2P(x \geq 9)$ where x is a binomial random variable with n = 10 and p = .5. From Table II,

$$\text{p-value} = 2P(x \geq 9) = 2(1 - P(x \leq 8)) = 2(1 - .989) = .022$$

Since the p-value = .022 $<$ α = .05, H_0 is rejected. There is sufficient evidence to indicate the median is different than 18 at α = .05.

e. For all parts, $\mu = np = 10(.5) = 5$ and $\sigma = \sqrt{npq} = \sqrt{10(.5)(.5)}$ = 1.581.

For part (a), $P(x \geq 6) \approx P\left[z > \frac{(6 - .5) - 5}{1.581}\right]$

$$= P(z \geq .32) = .5 - .1255 = .3745$$

This is close to the probability .377 in part (a). The conclusion is the same.

For part (b), $2P(x \geq 6) \approx 2P\left[z > \frac{(6 - .5) - 5}{1.581}\right]$

$$= 2P(z \geq .32) = 2(.5 - .1255) = .7890$$

This is close to the probability .754 in part (b). The conclusion is the same.

For part (c), $P(x \geq 9) \approx P\left[z > \frac{(9 - .5) - 5}{1.581}\right]$

$$= P(z \geq 2.21) = .5 - .4864 = .0136$$

This is close to the probability .011 in part (c). The conclusion is the same.

For part (d), $2P(x \geq 9) \approx 2P\left[z > \frac{(9 - .5) - 5}{1.581}\right]$

$$= 2P(z \geq 2.21) = 2(.5 - .4864) = .0272$$

This is close to the probability .022 in part (d). The conclusion is the same.

f. We must assume only that the sample is selected randomly from a continuous probability distribution.

7.51 a. I would recommend the sign test because 5 of the sample measurements are of similar magnitude, but the 6th is about three times as large as the others. It would be very unlikely to observe this sample if the population were normal.

b. To determine if the airline is meeting the requirement, we test:

H_0: $M = 30$
H_a: $M < 30$

c. The test statistic is S = {Number of measurements less than 30} = 5.

H_0 will be rejected if the p-value $< \alpha = .01$.

d. The test statistic is $S = 5$.

The p-value $= P(x \geq 5)$ where x is a binomial random variable with $n = 6$ and $p = .5$. From Table II,

$$\text{p-value} = P(x \geq 5) = 1 - P(x \leq 4) = 1 - .891 = .109$$

Since the p-value = .109 is not less than $\alpha = .01$, H_0 is not rejected. There is insufficient evidence to indicate the airline is meeting the maintenance requirement at $\alpha = .01$.

7.53 a. To determine if the median height exceeds 40 feet, we test:

H_0: $M = 40$
H_a: $M > 40$

The test statistic is S = {Number of measurements greater than 40} = 17.

The p-value $= P(x \geq 17)$ where x is a binomial random variable with $n = 24$ and $p = .5$. Since we do not have a table for $n = 24$ (and n is sufficiently large), we will use the large-sample approximation.

$$\mu = np = 24(.5) = 12 \text{ and } \sigma = \sqrt{npq} = \sqrt{24(.5)(.5)} = 2.4495$$

$$\text{p-value} = P(x \geq 17) \approx P\left(z \geq \frac{(17 - .5) - 12}{2.4495}\right)$$
$$= P(z \geq 1.84) = .5 - .4671 = .0329$$

Since the p-value = .0329 is less than $\alpha = .05$, H_0 is rejected. There is sufficient evidence to indicate the median height of the trees is more than 40 feet at $\alpha = .05$.

b. The p-value = .0329. If the true median of the trees is 40 feet, the probability of seeing 17 or more out of 24 trees taller than 40 feet is .0329.

c. We must assume the sample of 24 tree heights was randomly selected from a continuous probability distribution.

7.55 b. First, check to see if n is large enough.

$$p_0 \pm 3\sigma_{\hat{p}} \Rightarrow p_0 \pm 3\sqrt{p_0 q_0/n}$$

$$\Rightarrow .75 \pm 3\sqrt{(.75)(.25)/100}$$

$$\Rightarrow .75 \pm .13 \Rightarrow (.62, .88)$$

Since the interval lies within the interval (0, 1), the normal approximation will be adequate.

H_0: $p = .75$
H_a: $p < .75$

The test statistic is $z = \dfrac{\hat{p} - p_0}{\sigma_{\hat{p}}} = \dfrac{\hat{p} - p_0}{\sqrt{pq/n}} \approx \dfrac{\hat{p} - p_0}{\sqrt{p_0 q_0/n}}$

$$= \frac{.69 - .75}{\sqrt{.75(.25)/100}} = -1.39$$

The rejection region requires $\alpha = .05$ in the lower tail of the z distribution. From Table III, Appendix A, $z_{.05} = 1.645$. The rejection region is $z < -1.645$.

Since the observed value of the test statistic does not fall in the rejection region ($-1.39 \not< -1.645$), H_0 is not rejected. There is insufficient evidence to indicate that the proportion is less than .75 at $\alpha = .05$.

c. p-value $= P(z < -1.39) = .5 - P(0 < z < 1.39)$
$= .5 - .4177 = .0823$

7.57 From Exercise 6.44, n = 50 and since p is the proportion of consumers who do not like the snack food, $\hat{p}$ will be:

$$\hat{p} = \frac{\text{number of 0's in sample}}{n} = \frac{29}{50} = .58$$

First, check to see if the normal approximation will be adequate:

$$p_0 \pm 3\sigma_{\hat{p}} \Rightarrow p_0 \pm 3\sqrt{\frac{pq}{n}} \approx p_0 \pm 3\sqrt{\frac{p_0 q_0}{n}}$$

$$\Rightarrow .5 \pm 3\sqrt{\frac{.5(1 - .5)}{50}}$$

$$\Rightarrow .5 \pm .2121 \Rightarrow (.2879, .7121)$$

Since the interval lies completely in the interval (0, 1), the normal approximation will be adequate.

a. H_0: $p = .5$
H_a: $p > .5$

The test statistic is $z = \frac{\hat{p} - p_0}{\sigma_{\hat{p}}} = \frac{\hat{p} - p_0}{\sqrt{\frac{pq}{n}}} \approx \frac{\hat{p} - p_0}{\sqrt{\frac{p_0 q_0}{n}}}$

$$= \frac{.58 - .5}{\sqrt{\frac{.5(1 - .5)}{50}}} = 1.13$$

The rejection region requires $\alpha = .10$ in the upper tail of the z distribution. From Table III, Appendix A, $z_{.10} = 1.28$. The rejection region is $z > 1.28$.

Since the observed value of the test statistic does not fall in the rejection region ($z = 1.13 \not> 1.28$), H_0 is not rejected. There is insufficient evidence to indicate the proportion of customers who do not like the snack food is greater than .5 at $\alpha = .10$.

b. $\text{p-value} = P(z \geq 1.13) = .5 - P(0 < z < 1.13)$
$= .5 - .3708$
$= .1292$

7.59 a. The sample size is sufficiently large if $p_0 \pm 3\sigma_{\hat{p}}$ does not include 0 or 1.

$$p_0 \pm 3\sigma_{\hat{p}} \approx .65 \pm 3\sqrt{\frac{.65(.35)}{100}} \Rightarrow .65 \pm .143 \Rightarrow (.507, .793)$$

Since this interval does not contain 0 or 1, the sample size is sufficiently large.

b. To determine if the percentage of 1980 shoppers using cents-off coupons exceeds 65%, we test:

H_0: $p = .65$
H_a: $p > .65$

The test statistic is $z = \frac{\hat{p} - p_0}{\sqrt{\frac{p_0 q_0}{n}}} = \frac{.76 - .65}{\sqrt{\frac{.65(.35)}{100}}} = 2.31$

The rejection region requires $\alpha = .05$ in the upper tail of the z distribution. From Table III, Appendix A, $t_{.05} = 1.645$. The rejection region is $z > 1.645$.

Since the observed value of the test statistic falls in the rejection region ($z = 2.31 > 1.645$), H_0 is rejected. There is sufficient evidence to indicate the percentage of 1980 shoppers using cents-off coupons exceeds 65% at $\alpha = .05$.

c. The p-value $= P(z \geq 2.31) = .5 - .4896 = .0104$. If H_0 is true, the probability of observing our test statistic or anything more unusual is .0104.

7.61 a. Yes, it appears that more than 50% of the public has an unfavorable opinion since more than 50% of the sampled people felt this way.

b. Check to see if the normal approximation is appropriate.

$$p_0 \pm 3\sigma_{\hat{p}} \Rightarrow p_0 \pm 3\sqrt{\frac{pq}{n}} \Rightarrow p_0 \pm 3\sqrt{\frac{p_0 q_0}{n}}$$

$$\Rightarrow .5 \pm 3\sqrt{\frac{.5(.5)}{6788}}$$

$$\Rightarrow .5 \pm .0182 \Rightarrow (.4818, .5182)$$

This interval lies within the interval (0, 1), so the normal approximation is adequate.

To determine if more than 50% of the general public hold unfavorable opinions, we test:

H_0: $p = .5$
H_a: $p > .5$

The test statistic is $z = \dfrac{\hat{p} - p_0}{\sigma_{\hat{p}}} = \dfrac{\hat{p} - p_0}{\sqrt{\frac{pq}{n}}} \approx \dfrac{\hat{p} - p_0}{\sqrt{\frac{p_0 q_0}{n}}}$

$$= \frac{.51046 - .50}{\sqrt{\frac{.5(.5)}{6788}}} = 1.72 \quad \text{where } \hat{p} = \frac{3,465}{6,788} = .51046$$

The rejection region requires $\alpha = .05$ in the upper tail of the z distribution. From Table III, Appendix A, $z_{.05} = 1.645$. The rejection region is $z > 1.645$.

Since the observed value of the test statistic falls in the rejection region ($z = 1.72 > 1.645$), H_0 is rejected. There is sufficient evidence to indicate the proportion of the general public that hold unfavorable opinions is greater than .5 at $\alpha = .05$.

c. Check to see if the normal approximation will be adequate:

$$\hat{p} \pm 3\sigma_{\hat{p}} \Rightarrow \hat{p} \pm 3\sqrt{\frac{pq}{n}} \Rightarrow \hat{p} \pm 3\sqrt{\frac{p_0q_0}{n}} \text{ where } \hat{p} = \frac{821}{6,788} = .1209$$

$$\Rightarrow .1209 \pm 3\sqrt{\frac{.1209(.8791)}{6,788}}$$

$$\Rightarrow .1209 \pm .0119 \Rightarrow (.109, .1328)$$

Since the interval lies within in the interval (0, 1), the normal approximation will be adequate.

For confidence coefficient .90, $\alpha = .10$ and $\alpha/2 = .05$. From Table III, Appendix A, $z_{.05} = 1.645$. The 90% confidence interval is:

$$\hat{p} \pm z_{.05}\sqrt{\frac{pq}{n}}$$

$$\Rightarrow \hat{p} \pm 1.645\sqrt{\frac{\hat{p}\hat{q}}{n}}$$

$$\Rightarrow .1209 \pm 1.645\sqrt{\frac{.1209(.8791)}{6788}}$$

$$\Rightarrow .1209 \pm .0065 \Rightarrow (.1144, .1274)$$

d. The experiment must be binomial. Therefore, the people surveyed were independently chosen from the general public.

7.63 As β increases, the power $= 1 - \beta$ decreases. As β decreases, the power of the test increases.

7.65 a. From Exercise 7.64, we want to find $\bar{x}_0$ such that

$$P(\bar{x} > \bar{x}_0) = .05$$

$$P(\bar{x} > \bar{x}_0) = P\left(z > \frac{\bar{x}_0 - 1000}{120/\sqrt{36}}\right) = P\left(z > \frac{\bar{x}_0 - 1000}{20}\right) = .05$$

$P(z > z_0) = .05 \Rightarrow$ We want to find z_0 such that .05 of the area is to the right of it and $.5 - .05 = .4500$ of the area is between 0 and z_0. From Table III, Appendix A, $z_0 = 1.645$.

$$z_0 = \frac{\bar{x}_0 - 1000}{20} \Rightarrow 1.645 = \frac{\bar{x}_0 - 1000}{20} \Rightarrow \bar{x}_0 = 1032.9$$

If $\mu = 1040$, $\beta = P(\bar{x} < 1032.9) = P\left(z < \frac{1032.9 - 1040}{120/\sqrt{36}}\right)$

$$= P(z < -.36)$$

$$= .5 - .1406 = .3594$$

b. Power $= 1 - \beta = 1 - .3594 = .6406$

c. From Exercise 7.64, β = .74 and the power is .26.

The value of β when μ = 1040 is .3613 as compared with .74 when μ = 1020. Thus, the probability of accepting H_0 when it is false when μ = 1040 is smaller than when μ = 1020.

The power when μ = 1040 is .6387 as compared with .26 when μ = 1020. Thus, the probability of rejecting H_0 when it is false is larger when μ = 1040 than when μ = 1020.

7.67 From Exercise 7.66, we know $\mu_{\bar{x}} = \mu = 50$ and $\sigma_{\bar{x}} = \sigma/\sqrt{n} = 20/\sqrt{64} = 2.5$. The rejection region requires α = .10 in the lower tail of the z distribution. From Table III, Appendix A, $z_{.10} = 1.28$. The rejection region is $z < -1.28$.

$$\bar{x}_0 = \mu_0 - z_{.10}\sigma_{\bar{x}} = 50 - 1.28(2.5) = 46.8$$

For $\mu = 48$, $\beta = P(\bar{x} > 46.8) = P\left(z > \frac{46.8 - 48}{2.5}\right) = P(z > -.48)$

$= .5 + .1844 = .6844$

Power $= 1 - \beta = 1 - .6844 = .3156$.

From Exercise 7.66, Power = .7642. The power for Exercise 7.67 is less than that for Exercise 7.66.

7.69 a. For $\mu = 49$, $\beta = P(\bar{x} > 46.8) = P\left(z > \frac{46.8 - 49}{2.5}\right) = P(z > -.88)$

$= .5 + .3106 = .8106$ (from Table III, Appendix A)

For $\mu = 47$, $\beta = P(\bar{x} > 46.8) = P\left(z > \frac{46.8 - 47}{2.5}\right) = P(z > -.08)$

$= .5 + .0319 = .5319$

For $\mu = 45$, $\beta = P(\bar{x} > 46.8) = P\left(z > \frac{46.8 - 45}{2.5}\right) = P(z > .72)$

$= .5 - .2642 = .2358$

For $\mu = 43$, $\beta = P(\bar{x} > 46.8) = P\left(z > \frac{46.8 - 43}{2.5}\right) = P(z > 1.52)$

$= .5 -.4357 = .0643$

For $\mu = 41$, $\beta = P(\bar{x} > 46.8) = P\left(z > \frac{46.8 - 41}{2.5}\right) = P(z > 2.32)$

$= .5 -.4898 = .0102$

b.

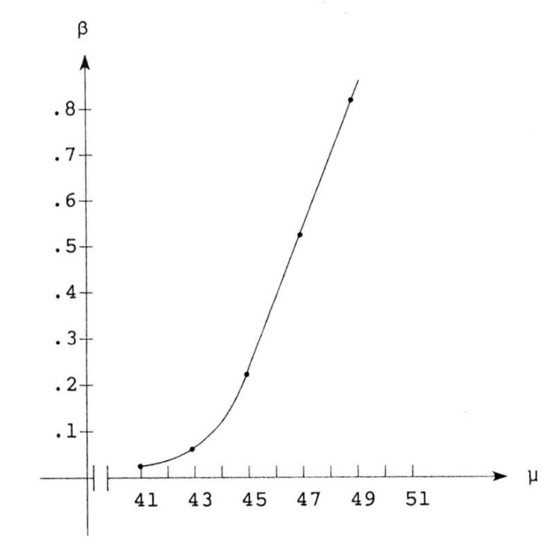

c. From the graph, when $\mu = 48$, $\beta \approx .67$. From Exercise 7.67, $\beta = .6844$.

d. For $\mu = 49$, Power $= 1 - \beta = 1 - .8106 = .1894$

For $\mu = 47$, Power $= 1 - \beta = 1 - .5319 = .4681$

For $\mu = 45$, Power $= 1 - \beta = 1 - .2358 = .7642$

For $\mu = 43$, Power $= 1 - \beta = 1 - .0643 = .9357$

For $\mu = 41$, Power $= 1 - \beta = 1 - .0102 = .9898$

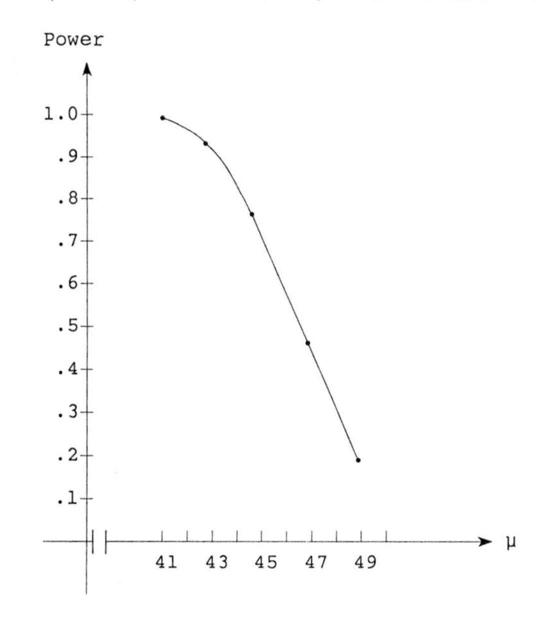

The power decreases as μ increases, while the value of β increases as μ increases (or gets closer to $\mu = 50$).

e. The further μ is from $\mu_0 = 50$, the larger the power and the smaller the value of β. This indicates the probability of rejecting H_0 when μ gets further from μ_0 (power) increases. The probability of accepting H_0 when μ gets further from μ_0 (β) decreases. The further μ gets from μ_0, the smaller the chances of making an incorrect decision.

7.71 a. From Exercise 7.14, $\alpha = .05$. The rejection region requires $\alpha = .05$ in the upper tail of the z distribution. From Table III, Appendix A, $z_{.05} = 1.645$.

$$\bar{x}_0 = \mu_0 - z_{.05}\sigma_{\bar{x}} = 35 + 1.645(6/\sqrt{36}) = 36.645$$

Power $= 1 - \beta = 1 - P(\bar{x} < 36.645)$ for given values of $\mu > 35$.

For $\mu = 35.5$, Power $= 1 - P(\bar{x} < 36.645) = 1 - P\left(z < \frac{36.645 - 35.5}{6/\sqrt{36}}\right)$

$= 1 - P(z < 1.15) = 1 - (.5 + .3749) = .1251$

For $\mu = 36.0$, Power $= 1 - P(\bar{x} < 36.645) = 1 - P\left(z < \frac{36.645 - 36}{1}\right)$

$= 1 - P(z < .65) = 1 - (.5 + .2422) = .2578$

For $\mu = 36.5$, Power $= 1 - P(\bar{x} < 36.645) = 1 - P\left(z < \frac{36.645 - 36.5}{1}\right)$

$= 1 - P(z < .15) = 1 - (.5 + .0596) = .4404$

For $\mu = 37.0$, Power $= 1 - P(\bar{x} < 36.645) = 1 - P\left(z < \frac{36.645 - 37}{1}\right)$

$= 1 - P(z < -.35) = 1 - (.5 + .1368) = .6368$

For $\mu = 37.5$, Power $= 1 - P(\bar{x} < 36.645) = 1 - P\left(z < \frac{36.645 - 37.5}{1}\right)$

$= 1 - P(z < -.85) = 1 - (.5 + .3023) = .8023$

b.

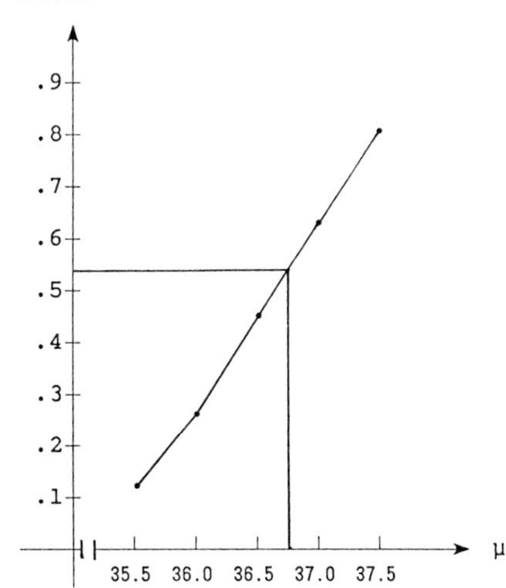

c. For $\mu = 36.75$, Power $\approx .54$

For $\mu = 36.75$, Power $= 1 - P(\bar{x} < 36.645)$

$= 1 - P\left(z < \frac{36.645 - 36.75}{1}\right)$

$= 1 - P(z < -.11) = 1 - (.5 - .0438) = .5438$

d. For $\mu = 40$, $\beta = P(\bar{x} < 36.645) = P\left(z < \frac{36.645 - 40}{1}\right)$

$= P(z < -3.36) \approx .5 - .5 = 0$

Power $= 1 - \beta \approx 1 - 0 = 1$

If $\mu = 40$, the chance that the test will fail to reject the null hypothesis is approximately 0.

7.73 a. The rejection region requires $\alpha = .05$ in the lower tail of the z distribution. From Table III, Appendix A, $z_{.05} = 1.645$. The rejection region is $z < -1.645$.

$\bar{x}_0 = \mu_0 - z_{.05}\sigma_{\bar{x}} = 5.0 - 1.645(.01/.\sqrt{100}) = 4.998355$

For $\mu = 4.9975$, $\beta = P(\bar{x} > 4.998355) = P\left(z > \frac{4.998355 - 4.9975}{.01/\sqrt{100}}\right)$

$= P(z > .855) = .5 - \left(\frac{.3023 + .3051}{2}\right) = .1963$

Failing to reject H_0 when H_0 is false is a Type II error.

b. For $\mu = 5.0$, $\alpha = P(\bar{x} < 4.998355) = P\left(z < \frac{4.998355 - 5}{.01/\sqrt{100}}\right)$

$= P(z < -1.645) = .5 - .4500 = .05$

Rejecting H_0 when H_0 is true is a Type I error.

c. .0025 mm below $\mu_0 = 5.0$ is $\mu = 5 - .0025 = 4.9975$.

For $\mu = 4.9975$, Power $= 1 - \beta = 1 - P(\bar{x} > 4.998355)$

$= 1 - P\left(z > \frac{4.998355 - 4.9975}{.01/\sqrt{100}}\right)$

$= 1 - P(z > .855)$

$= 1 - \left(.5 - \frac{.3023 + .3051}{2}\right)$

$= 1 - .1963 = .8037$

7.75 The elements of the test of hypothesis that should be specified prior to analyzing the data are: null hypothesis, alternative hypothesis, and rejection region.

7.77 For a large sample test of hypothesis about a population mean, no assumptions are necessary because the Central Limit Theorem assures that the test statistic will be approximately normally distributed. For a small sample test of hypothesis about a population mean, we must assume that the population being sampled from is normal. The test statistic for the large sample test is the z statistic, and the test statistic for the small sample test is the t statistic.

7.79 a. $H_0: \mu = 80$
$H_a: \mu < 80$

The test statistic is $t = \dfrac{\bar{x} - \mu_0}{s/\sqrt{n}} = \dfrac{72.6 - 80}{\sqrt{19.4}/\sqrt{20}} = -7.51$

The rejection region requires $\alpha = .05$ in the lower tail of the t distribution with df = n - 1 = 20 - 1 = 19. From Table IV, Appendix A, $t_{.05} = 1.734$. The rejection region is $t < -1.734$.

Since the observed value of the test statistic falls in the rejection region ($-7.51 < -1.734$), H_0 is rejected. There is sufficient evidence to indicate that the mean is less than 80 at $\alpha = .05$.

b. $H_0: \mu = 80$
$H_a: \mu \neq 80$

The test statistic is $t = \dfrac{\bar{x} - \mu_0}{s/\sqrt{n}} = \dfrac{72.6 - 80}{\sqrt{19.4}/\sqrt{20}} = -7.51$

The rejection region requires $\alpha/2 = .01/2 = .005$ in each tail of the t distribution with df = n - 1 = 20 - 1 = 19. From Table IV, Appendix A, $t_{.005} = 2.861$. The rejection region is $z < -2.861$ or $z > 2.861$.

Since the observed value of the test statistic falls in the rejection region ($-7.51 < -2.861$), H_0 is rejected. There is sufficient evidence to indicate that the mean is other than 80 at $\alpha = .01$.

7.81 a. $H_0: \mu = 8.3$
$H_a: \mu \neq 8.3$

The test statistic is $z = \dfrac{\bar{x} - \mu_0}{\sigma_{\bar{x}}} = \dfrac{8.2 - 8.3}{.79/\sqrt{175}} = -1.67$

The rejection region requires $\alpha/2 = .05/2 = .025$ in each tail of the z distribution. From Table III, Appendix A, $z_{.025} = 1.96$. The rejection region is $z < -1.96$ or $z > 1.96$.

Since the observed value of the test statistic does not fall in the rejection region ($-1.67 \nless -1.96$), H_0 is not rejected. There is insufficient evidence to indicate that the mean is different from 8.3 at $\alpha = .05$.

b. $H_0: \mu = 8.4$
$H_a: \mu \neq 8.4$

The test statistic is $z = \dfrac{\bar{x} - \mu_0}{\sigma_{\bar{x}}} = \dfrac{8.2 - 8.4}{.79/\sqrt{175}} = -3.35$

The rejection region is the same as part (b), $z < -1.96$ or $z > 1.96$.

Since the observed value of the test statistic falls in the rejection region ($-3.35 < -1.96$), H_0 is rejected. There is sufficient evidence to indicate that the mean is different from 8.4 at $\alpha = .05$.

7.83 a. First, check to see if n is large enough.

$$p_0 \pm 3\sigma_{\hat{p}} \Rightarrow p_0 \pm 3\sqrt{p_0 q_0/n}$$
$$\Rightarrow .50 \pm 3\sqrt{(.50)(.50)/350}$$
$$\Rightarrow .50 \pm .08 \Rightarrow (.42, .58)$$

Since the interval lies within the interval (0, 1), the normal approximation will be adequate.

H_0: $p = .50$
H_a: $p \neq .50$

The test statistic is $z = \dfrac{\hat{p} - p_0}{\sigma_{\hat{p}}} = \dfrac{\hat{p} - p_0}{\sqrt{pq/n}} \approx \dfrac{\hat{p} - p_0}{\sqrt{p_0 q_0/n}}$

$$= \frac{.42 - .50}{\sqrt{.50(.50)/350}} = -2.99$$

The rejection region requires $\alpha/2 = .01/2 = .005$ in each tail of the z distribution. From Table III, Appendix A, $z_{.005} = 2.58$. The rejection region is $z < -2.58$ or $z > 2.58$.

Since the observed value of the test statistic falls in the rejection region ($-2.99 < -2.58$), H_0 is rejected. There is sufficient evidence to indicate that the proportion of loans to students in this one region is different from .50 at $\alpha = .01$.

b. p-value $= 2P(z < -2.99) = 2\{.5 - P(0 < z < 1.39)\}$
$= 2(.5 - .4986) = .0028$

7.85 a. To determine if the individual is in the normal range, we test:

H_0: $M = 1$
H_a: $M < 1$

b. The test statistic is S = {Number of observations less than 1} = 7

The p-value $= P(x \geq 7)$ where x is a binomial random variable with $n = 9$ and $p = .5$. From Table II,

p-value $= P(x \geq 7) = 1 - P(x \leq 6) = 1 - .910 = .090$

Since the p-value = .090 > α = .05, H_0 is not rejected. There is insufficient evidence to indicate the individual is in the normal range at α = 05.

c. We must assume the sample is a random sample from a continuous distribution.

7.87 a. H_0: $\mu = 3$
H_a: $\mu > 3$

where μ = mean amount of PCB in the effluent.

The test statistic is $z = \frac{\bar{x} - \mu_0}{\sigma_{\bar{x}}} = \frac{\bar{x} - \mu_0}{\sigma/\sqrt{n}} \approx \frac{\bar{x} - \mu_0}{s/\sqrt{n}}$

$$= \frac{3.1 - 3}{.5/\sqrt{50}}$$

$$= 1.41$$

The rejection region requires α =.01 in the upper tail of the z distribution. From Table III, Appendix A, $z_{.01}$ = 2.33. The rejection region is z > 2.33.

Since the observed value of the test statistic does not fall in the rejection region, ($z = 1.41 \not> 2.33$), H_0 is not rejected. There is insufficient evidence to indicate the mean amount of PCB in the effluent is more than 3 parts per million at α = .01. Do not halt the manufacturing process.

b. As plant manager, I do not want to shut down the plant unnecessarily. Therefore, I want α = P(shut down plant when μ = 3) to be small.

7.89 a. No, it increases the risk of falsely rejecting H_0, i.e., closing the plant unnecessarily.

b. First, find

$\bar{x}_0 = \mu_0 + z_\alpha \sigma_{\bar{x}} \approx \mu_0 + z_{.05}\frac{s}{\sqrt{n}}$ where $z_{.05}$ = 1.645 from Table III, Appendix A.

$$\bar{x} = 3 + 1.645\frac{.5}{\sqrt{50}}$$

$$= 3.116$$

Then, compute

$$\beta = P(\bar{x}_0 < 3.116 \text{ when } \mu = 3.1)$$

$$z = \frac{\bar{x}_0 - \mu_a}{\sigma_{\bar{x}}} \approx \frac{\bar{x}_0 - \mu_a}{s/\sqrt{n}}$$

$$= \frac{3.116 - 3.1}{.5/\sqrt{50}}$$

$$= .23$$

$$\beta = P(z < .23) = .5 + P(0 < z < .23)$$

$$= .5 + .0910$$

$$= .5910$$

$$\text{power} = 1 - \beta = 1 - .5910 = .4090$$

c. The power of the test increases as α increases.

7.91 To determine if the median exceeds 8 pounds, we test:

H_0: $M = 8$
H_a: $M > 8$

The test statistic is S = {Number of observations greater than 8} = 5.

The p-value = $P(x \geq 5)$ where x is a binomial random variable with $n = 6$ and $p = .5$. From Table II,

$$\text{p-value} = P(x \geq 5) = 1 - P(x \leq 4) = 1 - .891 = .109$$

Since the p-value = $.109 > \alpha = .05$, H_0 is not rejected. There is insufficient evidence to indicate the median exceeds 8 pounds at $\alpha = .05$.

7.93 Using Table III, Appendix A, the p-value is $P(z > 1.41) = .5 - .4207 = .0793$. The probability of observing a test statistic of 1.41 or anything more unusual is .0793. This is not particularly small. There is no evidence to reject H_0.

7.95 To determine if the proportion of defective can openers in the shipment exceeds .02, we test:

H_0: $p = .02$
H_a: $p > .02$

First, check to see if the normal approximation will be adequate:

$$p_0 \pm 3\sigma_{\hat{p}} = p_0 \pm 3\sqrt{\frac{pq}{n}} \approx p_0 \pm 3\sqrt{\frac{p_0q_0}{n}}$$

$$\Rightarrow .02 \pm 3\sqrt{\frac{.02(.98)}{400}}$$

$$\Rightarrow .02 \pm .021 \Rightarrow (-.001, .041)$$

Since the interval does not lie completely in the interval (0, 1), the normal approximation will not be adequate.

We will use the approximation to get an idea of the answer.

The test statistic is $z = \frac{\hat{p} - p_0}{\sigma_{\hat{p}}} = \frac{\hat{p} - p_0}{\sqrt{\frac{pq}{n}}} \approx \frac{\hat{p} - p_0}{\sqrt{\frac{p_0q_0}{n}}}$

$$= \frac{.0275 - .02}{\sqrt{\frac{.02(.98)}{400}}} = 1.07, \text{ where } \hat{p} = \frac{11}{400} = .0275$$

$$\text{p-value} = P(z \geq 1.07) = .5 - P(0 < z < 1.07)$$
$$= .5 - .3577$$
$$= .1423$$

Reject H_0 if p-value $< \alpha$
$.1423 \nless .05$

Fail to reject H_0. There is insufficient evidence to indicate the proportion of defective can openers exceeds .02 at $\alpha = .05$.

7.97 For confidence coefficient .95, $\alpha = 1 - .95 = .05$ and $\alpha/2 = .05/2 = .025$. From Table III, Appendix A, $z_{.025} = 1.96$.

$$n = \frac{(z_{\alpha/2})\sigma^2}{B^2} \approx \frac{(1.96)^2(2.5)^2}{(.45)^2} = 118.6 \approx 119$$

7.99 a. The value of the test statistic is $t = 2.408$. The p-value is .0304, which corresponds to a two-tailed test. The p-value = $P(t \geq 2.408) + P(t \leq -2.408) = .0304$. Since the p-value is less than $\alpha = .10$, H_0 is rejected. There is sufficient evidence to indicate the mean beta coefficient of high technology stock is different than 1.

b. The p-value would be $.0304/2 = .0152$.

CHAPTER

8

INFERENCES FOR MEANS BASED ON TWO SAMPLES

8.1 a. $\mu_1 \pm 2\sigma_{\bar{x}_1}$

$\Rightarrow \mu_1 \pm 2\frac{\sigma_1}{\sqrt{n_1}}$

$\Rightarrow 150 \pm 2\frac{\sqrt{900}}{\sqrt{100}}$

$\Rightarrow 150 \pm 6$

$\Rightarrow (144,\ 156)$

b. $\mu_2 \pm 2\sigma_{\bar{x}_2}$

$\Rightarrow \mu_2 \pm 2\ \frac{\sigma_2}{\sqrt{n_2}}$

$\Rightarrow 150 \pm 2\ \frac{\sqrt{1600}}{\sqrt{100}}$

$\Rightarrow 150 \pm 8$

$\Rightarrow (142,\ 158)$

c. $\mu_1 - \mu_2 = 150 - 150 = 0$

$\sigma_{\bar{x}_1-\bar{x}_2} = \sqrt{\frac{\sigma_1^2}{n_1} + \frac{\sigma_2^2}{n_2}}$

$= \sqrt{\frac{900}{100} + \frac{1600}{100}}$

$= \sqrt{\frac{2500}{100}}$

$= 5$

d. $(\mu_1 - \mu_2) \pm 2\ \sqrt{\frac{\sigma_1^2}{n_1} + \frac{\sigma_2^2}{n_2}}$

$\Rightarrow (150 - 150) \pm 2\ \sqrt{\frac{900}{100} + \frac{1600}{100}}$

$\Rightarrow 0 \pm 10$

$\Rightarrow (-10,\ 10)$

e. The variability of the difference between the sample means is greater than the variability of the individual sample means.

8.3 a. $z = \frac{(\bar{x}_1 - \bar{x}_2) - (\mu_1 - \mu_2)}{\sqrt{\frac{\sigma_1^2}{n_1} + \frac{\sigma_2^2}{n_2}}} = \frac{1.5 - (12 - 10)}{\sqrt{\frac{4^2}{64} + \frac{3^2}{64}}} = -.8$

b. $P(\bar{x}_1 - \bar{x}_2 > 1.5) = P(z > -.8) = .5 + .2881 = .7881$ (from Table III, Appendix A.

c. $P(\bar{x}_1 - \bar{x}_2 < 1.5) = P(z < -.8) = .5 - .2881 = .2119$

d. $P(\bar{x}_1 - \bar{x}_2 < -1) + P(\bar{x}_1 - \bar{x}_2 > 1)$

$$= P\left[z < \frac{-1 - (12 - 10)}{\sqrt{\frac{4^2}{64} + \frac{3^2}{64}}}\right] + P\left[z > \frac{1 - (12 - 10)}{\sqrt{\frac{4^2}{64} + \frac{3^2}{64}}}\right]$$

$$= P(z < -4.8) + P(z > -1.6) \approx .5 - .5 + .5 + .4452 = .9452$$

8.5 a. The sampling distribution of $\bar{x}_1 - \bar{x}_2$ is approximately normal by the Central Limit Theorem since $n_1 \geq 30$ and $n_2 \geq 30$.

$$\mu_{\bar{x}_1 - \bar{x}_2} = \mu_1 - \mu_2 = -5$$

$$\sigma_{\bar{x}_1 - \bar{x}_2} = \sqrt{\frac{\sigma_1^2}{n_1} + \frac{\sigma_2^2}{n_2}} \approx \sqrt{\frac{s_1^2}{n_1} + \frac{s_2^2}{n_2}} = \sqrt{\frac{10^2}{100} + \frac{12^2}{150}} = \sqrt{1.96} = 1.4$$

b. H_0: $\mu_1 - \mu_2 = -5$
H_a: $\mu_1 - \mu_2 < -5$

The test statistic is $z = \frac{(\bar{x}_1 - \bar{x}_2) - (-5)}{\sqrt{\frac{\sigma_1^2}{n_1} + \frac{\sigma_2^2}{n_2}}} \approx \frac{(1025 - 1039) - (-5)}{1.4}$

$= -6.43$

The rejection region requires $\alpha = .01$ in the lower tail of the z distribution. From Table III, Appendix A, $z_{.01} = 2.33$. The rejection region is $z < -2.33$.

Since the observed value of the test statistic falls in the rejection region ($z = -6.43 < -2.33$), H_0 is rejected. There is sufficient evidence to indicate the difference in the population means is less than -5 at $\alpha = .01$.

c. p-value $= P(z \leq -6.43) = .5 - P(0 < z < 6.43)$
$\approx .5 - .5$
$= 0$

8.7 a. Both of the populations in the comparison are from a university. One of the populations consists of the number of study hours expended per week by student athletes and the other one consists of the number of study hours expended per week by nonathletes.

b. Let μ_1 = mean number of study hours expended per week by student athletes and μ_2 = mean number of study hours expended per week by nonathletes. To determine if there is a difference in the mean number of hours of study per week between athletes and nonathletes, we test:

H_0: $\mu_1 - \mu_2 = 0$
H_a: $\mu_1 - \mu_2 \neq 0$

The test statistic is $z = \frac{(\bar{x}_1 - \bar{x}_2) - D_0}{\sqrt{\frac{\sigma_1^2}{n_1} + \frac{\sigma_2^2}{n_2}}} \approx \frac{(\bar{x}_1 - \bar{x}_2) - D_0}{\sqrt{\frac{s_1^2}{n_1} + \frac{s_2^2}{n_2}}}$

$$= \frac{(20.6 - 23.5) - 0}{\sqrt{\frac{5.3^2}{55} + \frac{4.1^2}{200}}} = \frac{-2.9}{.7712} = -3.76$$

The rejection region requires $\alpha/2 = .01/2 = .005$ in each tail of the z distribution. From Table III, Appendix A, $z_{.005} \approx 2.58$. The rejection region is $z < -2.58$ or $z > 2.58$.

Since the test statistic falls in the rejection region ($z = -3.76 < -2.58$), H_0 is rejected. There is sufficient evidence to conclude that there is a difference in the mean number of hours of study per week between athletes and nonathletes at $\alpha = .01$.

c. For confidence coefficient .99, $\alpha = .01$ and $\alpha/2 = .005$. From Table III, Appendix A, $z_{.005} \approx 2.58$. The confidence interval is:

$$(\bar{x}_1 - \bar{x}_2) \pm z_{.005}\sqrt{\frac{\sigma_1^2}{n_1} + \frac{\sigma_2^2}{n_2}}$$

$$\Rightarrow (20.6 - 23.5) \pm 2.58\sqrt{\frac{5.3^2}{55} + \frac{4.1^2}{200}}$$

$$\Rightarrow -2.9 \pm 1.99 \Rightarrow (-4.89, -.91)$$

d. A 95% confidence interval would be narrower than a 99% confidence interval. The more the confidence, the wider the interval. For a 95% confidence interval, $z_{.025} = 1.96$ as opposed to $z_{.005} = 2.58$ for the 99% confidence interval.

8.9 For confidence coefficient .99, $\alpha = .01$ and $\alpha/2 = .005$. From Table III, Appendix A, $z_{.005} \approx 2.58$.

The 99% confidence interval for $\mu_1 - \mu_2$ is:

$$(\bar{x}_1 - \bar{x}_2) \pm z_{\alpha/2}\sqrt{\frac{\sigma_1^2}{n_1} + \frac{\sigma_2^2}{n_2}}$$

Estimating σ_1^2 and σ_2^2 with s_1^2 and s_2^2, we get

$$\Rightarrow (78.7 - 75.3) \pm 2.58\sqrt{\frac{201.6}{40} + \frac{259.2}{40}}$$

$$\Rightarrow 3.4 \pm 8.76 \Rightarrow (-5.36, 12.16)$$

8.11 a. To determine if the merged firms generally have smaller price earnings ratios, we test:

H_0: $\mu_1 = \mu_2$
H_a: $\mu_1 < \mu_2$
where μ_1 = mean price earnings ratio for merged firms and
μ_2 = mean price earnings ratio for nonmerged firms

The test statistic is $z = \dfrac{(\bar{x}_1 - \bar{x}_2) - 0}{\sqrt{\dfrac{\sigma_1^2}{n_1} + \dfrac{\sigma_2^2}{n_2}}} = \dfrac{(7.295 - 14.666) - 0}{\sqrt{\dfrac{7.374^2}{44} + \dfrac{16.089^2}{44}}}$

$= \dfrac{-7.371}{2.6681} = -2.76$

The rejection region requires $\alpha = .05$ in the lower tail of the z distribution. From Table III, Appendix A, $z_{.05} = 1.645$. The rejection region is $z < -1.645$.

Since the observed value of the test statistic falls in the rejection region ($z = -2.76 < -1.645$), H_0 is rejected. There is sufficient evidence to indicate the mean price earnings ratio for merged firms is smaller than that for nonmerged firms at $\alpha = 05$.

b. The p-value is $P(z \leq -2.76) = .5 - .4971 = .0029$.

c. We must assume the samples are independently and randomly selected.

d. No. If the price-earnings ratios cannot be negative, the populations cannot be normal because the standard deviations for both groups are larger than the means. Thus, there can be no observations more than one standard deviation below the mean.

8.13 a. Let μ_1 = mean initial performance measure of stayers and μ_2 = mean initial performance measure of leavers. Since we want to test the null hypothesis that there is no difference in mean initial performance measure for stayers and leavers, we test:

H_0: $\mu_1 - \mu_2 = 0$
H_a: $\mu_1 - \mu_2 \neq 0$

The test statistic is $z = \dfrac{(\bar{x}_1 - \bar{x}_2) - D_0}{\sqrt{\dfrac{\sigma_1^2}{n_1} + \dfrac{\sigma_2^2}{n_2}}} \approx \dfrac{(\bar{x}_1 - \bar{x}_2) - D_0}{\sqrt{\dfrac{s_1^2}{n_1} + \dfrac{s_2^2}{n_2}}}$

$= \dfrac{(3.51 - 3.24) - 0}{\sqrt{\dfrac{.51^2}{174} + \dfrac{.52^2}{355}}} = \dfrac{.27}{.0475} = 5.68$

The rejection region requires $\alpha/2 = .01/2 = .005$ in each tail of the z distribution. From Table III, Appendix A, $z_{.005} \approx 2.58$. The rejection region is $z < -2.58$ or $z > 2.58$.

Since the test statistic falls in the rejection region ($z = 5.68 > 2.58$), H_0 is rejected. There is sufficient evidence to conclude that a difference does exist between the mean initial performance measure for stayers and leavers at $\alpha = .01$.

b. The p-value $= P(z \leq -5.68) + P(z \geq 5.68) \approx (.5 - .5) + (.5 - .5) = 0$.

Observing a value of z as large as 5.68 is an improbable event if, in fact, $\mu_1 = \mu_2$. Since the probability of this occurring is approximately zero, we would conclude that there is strong evidence to suggest $\mu_1 \neq \mu_2$.

8.15 a. Let μ_1 = mean final performance measure of stayers and μ_2 = mean final performance measure of leavers. Since we want to test the null hypothesis that there is no difference in mean final performance measure for stayers and leavers, we test:

$$H_0: \mu_1 - \mu_2 = 0$$
$$H_a: \mu_1 - \mu_2 \neq 0$$

The test statistic is $z = \dfrac{(\bar{x}_1 - \bar{x}_2) - D_0}{\sqrt{\dfrac{\sigma_1^2}{n_1} + \dfrac{\sigma_2^2}{n_2}}} \approx \dfrac{(\bar{x}_1 - \bar{x}_2) - D_0}{\sqrt{\dfrac{s_1^2}{n_1} + \dfrac{s_2^2}{n_2}}}$

$$= \frac{(3.78 - 3.15) - 0}{\sqrt{\dfrac{.62^2}{174} + \dfrac{.68^2}{355}}} = \frac{.63}{.05926} = 10.63$$

The rejection region requires $\alpha/2 = .01/2 = .005$ in each tail of the z distribution. From Table III, Appendix A, $z_{.005} \approx 2.58$. The rejection region is $z < -2.58$ or $z > 2.58$.

Since the test statistic falls in the rejection region ($z = 10.63 > 2.58$), H_0 is rejected. There is sufficient evidence to conclude that a difference does exist between the mean final performance measure for stayers and leavers at $\alpha = .01$.

b. The p-value $= P(z \leq -10.63) + P(z \geq 10.63) \approx (.5 - .5) + (.5 - .5) = 0$.

Observing a value z as large as 10.63 is an improbable event if, in fact, $\mu_1 = \mu_2$. Since the probability of this occurring is approximately zero, we would conclude that there is evidence to suggest $\mu_1 \neq \mu_2$.

8.17 Assumptions about the two populations:

1. Both sampled populations have relative frequency distributions that are approximately normal.
2. The population variances are equal.

Assumptions about the two samples:

The samples are randomly and independently selected from the population.

8.19 Since we are assuming $\sigma_1^2 = \sigma_2^2 = \sigma^2$, we have 2 estimators ($s_1^2$ and s_2^2) for the same parameter σ^2. Thus, we use the pooled estimator or weighted average of s_1^2 and s_2^2 to come up with one "good" estimator for σ^2 rather than 2 estimators.

8.21 Some preliminary calculations are:

$$\bar{x}_1 = \frac{\sum x_1}{n_1} = \frac{11.8}{5} = 2.36 \qquad s_1^2 = \frac{\sum x_1^2 - \frac{(\sum x_1)^2}{n_1}}{n_1 - 1} = \frac{30.78 - \frac{(11.8)^2}{5}}{5 - 1} = .733$$

$$\bar{x}_2 = \frac{\sum x_2}{n_2} = \frac{14.4}{4} = 3.6 \qquad s_2^2 = \frac{\sum x_2^2 - \frac{(\sum x_2)^2}{n_2}}{n_2 - 1} = \frac{53.1 - \frac{(14.4)^2}{4}}{4 - 1} = .42$$

a. $$s_p^2 = \frac{(n_1 - 1)s_1^2 + (n_2 - 1)s_2^2}{n_1 + n_2 - 2} = \frac{(5 - 1).733 + (4 - 1).42}{5 + 4 - 2} = \frac{4.192}{7}$$

$$= .5989$$

b. $H_0: \mu_1 - \mu_2 = 0$
$H_a: \mu_1 - \mu_2 < 0$

The test statistic is $$t = \frac{(\bar{x}_1 - \bar{x}_2) - D_0}{\sqrt{s_p^2\left(\frac{1}{n_1} + \frac{1}{n_2}\right)}} = \frac{(2.36 - 3.6) - 0}{\sqrt{.5989\left(\frac{1}{5} + \frac{1}{4}\right)}}$$

$$= \frac{-1.24}{.5191} = -2.39$$

The rejection region requires $\alpha = .10$ in the lower tail of the t distribution with df $= n_1 + n_2 - 2 = 5 + 4 - 2 = 7$. From Table IV, Appendix A, $t_{.10} = 1.415$. The rejection region is $t < -1.415$.

Since the test statistic falls in the rejection region ($t = -2.39 < -1.415$), H_0 is rejected. There is sufficient evidence to indicate that $\mu_2 > \mu_1$ at $\alpha = .10$.

c. A small sample confidence interval is needed because $n_1 = 5 < 30$ and $n_2 = 4 < 30$.

For confidence coefficient .90, $\alpha = .10$ and $\alpha/2 = .05$. From Table IV, Appendix A, with df $= n_1 + n_2 - 2 = 5 + 4 - 2 = 7$, $t_{.05} = 1.895$. The 90% confidence interval for $(\mu_1 - \mu_2)$ is:

$$(\bar{x}_1 - \bar{x}_2) \pm t_{.05}\sqrt{s_p^2\left(\frac{1}{n_1} + \frac{1}{n_2}\right)}$$

$$\Rightarrow (2.36 - 3.6) \pm 1.895\sqrt{.5989\left(\frac{1}{5} + \frac{1}{4}\right)}$$

$$\Rightarrow -1.24 \pm .98$$

$$\Rightarrow (-2.22, -0.26)$$

d. The confidence interval in part (c) provides more information about $(\mu_1 - \mu_2)$ than the test of hypothesis in part (b). The test in part (b) only tells us that μ_2 is greater than μ_1. However, the confidence interval estimates what the difference is between μ_1 and μ_2.

8.23 Some preliminary calculations:

$$\bar{x}_1 = \frac{\sum x_1}{n_1} = \frac{654}{15} = 43.6 \qquad s_1^2 = \frac{\sum x_1^2 - \frac{(\sum x_1)^2}{n_1}}{n_1 - 1} = \frac{28934 - \frac{654^2}{15}}{15 - 1} = \frac{419.6}{14}$$

$$= 29.9714$$

$$\bar{x}_2 = \frac{\sum x_2}{n_2} = \frac{858}{16} = 53.62 \qquad s_2^2 = \frac{\sum x_2^2 - \frac{(\sum x_2)^2}{n_2}}{n_2 - 1} = \frac{46450 - \frac{858^2}{16}}{16 - 1} = \frac{439.75}{15}$$

$$= 29.3167$$

$$s_p^2 = \frac{(n_1 - 1)s_1^2 + (n_2 - 1)s_2^2}{n_1 + n_2 - 2} = \frac{(15 - 1)29.9714 + (16 - 1)29.3167}{15 + 16 - 2}$$

$$= \frac{859.3501}{29} = 29.6328$$

a. $H_0\colon \mu_2 - \mu_1 = 10$
$H_a\colon \mu_2 - \mu_1 > 10$

The test statistic is $t = \frac{(\bar{x}_2 - \bar{x}_1) - D_0}{\sqrt{s_p^2\left(\frac{1}{n_1} + \frac{1}{n_2}\right)}} = \frac{(53.625 - 43.6) - 10}{\sqrt{29.6328\left(\frac{1}{15} + \frac{1}{16}\right)}}$

$$= \frac{.025}{1.9564} = .013$$

The rejection region requires $\alpha = .01$ in the upper tail of the t distribution with df $= n_1 + n_2 - 2 = 15 + 16 - 2 = 29$. From Table IV, Appendix A, $t_{.01} = 2.462$. The rejection region is $t > 2.462$.

Since the test statistic does not fall in the rejection region ($t = .013 \not> 2.462$), H_0 is not rejected. There is insufficient evidence to conclude $\mu_2 - \mu_1 > 10$ at $\alpha = .01$.

b. A small sample confidence interval is needed because $n_1 = 15 < 30$ and $n_2 = 16 < 30$.

For confidence coefficient .98, $\alpha = .02$ and $\alpha/2 = .01$. From Table IV, Appendix A, with df $= n_1 + n_2 - 2 = 15 + 16 - 2 = 29$, $t_{.01} = 2.462$. The 98% confidence interval for $(\mu_2 - \mu_1)$ is:

$$(\bar{x}_2 - \bar{x}_1) \pm t_{\alpha/2} \sqrt{s_p^2\left(\frac{1}{n_1} + \frac{1}{n_2}\right)}$$

$$\Rightarrow (53.625 - 43.6) \pm 2.462\sqrt{29.6328\left(\frac{1}{15} + \frac{1}{16}\right)}$$

$$\Rightarrow 10.025 \pm 4.817$$

$$\Rightarrow (5.208,\ 14.842)$$

We are 98% confident that the difference between the mean of population 2 and the mean of population 1 is between 5.208 and 14.842.

8.25 a. Let μ_1 = mean length of time to complete the task for the control group and μ_2 = mean length of time to complete the task for the experimental group. Since we want to see if the mean time is less for those subjects who receive the drug (experimental group), we test:

$H_0: \mu_1 - \mu_2 = 0$
$H_a: \mu_1 - \mu_2 > 0$

The test statistic is $t = \dfrac{(\bar{x}_1 - \bar{x}_2) - D_0}{\sqrt{s_p^2\left(\frac{1}{n_1} + \frac{1}{n_2}\right)}}$

where $s_p^2 = \dfrac{s_1^2 + s_2^2}{2} = \dfrac{3.9 + 4.3}{2} = 4.1$ (since $n_1 = n_2$, s_p^2 is just the average of s_1^2 and s_2^2.)

$$t = \frac{(14.8 - 12.3) - 0}{\sqrt{4.1\left(\frac{1}{10} + \frac{1}{10}\right)}} = \frac{2.5}{.9055} = 2.76$$

The rejection region requires $\alpha = .10$ in the upper tail of the t distribution with df $= n_1 + n_2 - 2 = 10 + 10 - 2 = 18$. From Table IV, Appendix A, $t_{.10} = 1.330$. The rejection region is $t > 1.330$.

Since the test statistic falls in the rejection region ($t = 2.76 > 1.330$), H_0 is rejected. There is sufficient evidence to indicate that the mean time to complete the task is less for those subjects who receive the drug at $\alpha = .10$.

b. The p-value = $P(t \geq 2.76)$.

Using Table IV, Appendix A, with df = 18,

$$.005 < \text{p-value} < .01$$

Observing a value of t as large as 2.76 is an improbable event if, in fact, $\mu_1 = \mu_2$. Since the probability of this occurring is between .005 and .01, we would conclude that there is evidence to suggest $\mu_1 - \mu_2 > 0$.

8.27 Let μ_1 = mean score on the comprehension test for method 1 and μ_2 = mean score on the comprehension test for method 2. Since we want to see if there is a difference in the mean scores on the test for the two teaching methods, we test:

$H_0: \mu_1 - \mu_2 = 0$
$H_a: \mu_1 - \mu_2 \neq 0$

The test statistic is $t = \dfrac{(\bar{x}_1 - \bar{x}_2) - D_0}{\sqrt{s_p^2\left(\frac{1}{n_1} + \frac{1}{n_2}\right)}}$

where $s_p^2 = \dfrac{(n_1 - 1)s_1^2 + (n_2 - 1)s_2^2}{n_1 + n_2 - 2} = \dfrac{(11 - 1)52 + (14 - 1)71}{11 + 14 - 2}$

$$= \frac{1443}{23} = 62.7391$$

$$t = \frac{(64 - 69) - 0}{\sqrt{62.7391\left(\frac{1}{11} + \frac{1}{14}\right)}} = \frac{-5}{3.1914} = -1.57$$

The rejection region requires $\alpha/2 = .05/2 = .025$ in each tail of the t distribution with df = $n_1 + n_2 - 2 = 14 + 11 - 2 = 23$. From Table IV, Appendix A, $t_{.025} = 2.069$. The rejection region is $t < -2.069$ or $t > 2.069$.

Since the test statistic does not fall in the rejection region ($t = -1.57 \nless -2.069$), H_0 is not rejected. There is insufficient evidence to indicate a difference in the mean scores for the two teaching methods at $\alpha = .05$.

8.29 Some preliminary calculations:

$$\bar{x}_1 = \frac{\sum x_1}{n_1} = \frac{192.6}{6} = 32.1 \qquad s_1^2 = \frac{\sum x_1^2 - \frac{(\sum x_1)^2}{n_1}}{n_1 - 1} = \frac{6233.3 - \frac{192.6^2}{6}}{6 - 1} = 10.168$$

$$\bar{x}_2 = \frac{\sum x_2}{n_2} = \frac{177.7}{6} = 29.6167 \qquad s_2^2 = \frac{\sum x_2^2 - \frac{(\sum x_2)^2}{n_2}}{n_2 - 1} = \frac{5290.61 - \frac{177.7^2}{6}}{6 - 1}$$

$$= 5.5457$$

Since $n_1 = n_2$,

$$s_p^2 = \frac{s_1^2 + s_2^2}{2} = \frac{10.168 + 5.5457}{2} = 7.85685$$

a. The bacteria counts are probably normally distributed because each count is an average of 5 measurements from the same specimen.

b. Let μ_1 = mean of the bacteria count for the discharge and μ_2 = mean of the bacteria count upstream. Since we want to test if the mean of the bacteria count for the discharge exceeds the mean of the count upstream, we test

$H_0: \mu_1 - \mu_2 = 0$
$H_a: \mu_1 - \mu_2 > 0$

The test statistic is $t = \dfrac{(\bar{x}_1 - \bar{x}_2) - D_0}{\sqrt{s_p^2\left(\frac{1}{n_1} + \frac{1}{n_2}\right)}} = \dfrac{(32.1 - 29.6167) - 0}{\sqrt{7.85685\left(\frac{1}{6} + \frac{1}{6}\right)}}$

$= \dfrac{2.4833}{1.6183} = 1.53$

The rejection region requires $\alpha = .05$ in the upper tail of the t distribution with df $= n_1 + n_2 - 2 = 6 + 6 - 2 = 10$. From Table IV, Appendix A, $t_{.05} = 1.812$. The rejection region is $t > 1.812$.

Since the test statistic does not fall in the rejection region ($t = 1.53 \not> 1.812$), H_0 is not rejected. There is insufficient evidence to indicate that the mean of the bacteria count for the discharge exceeds the mean of the count upstream at $\alpha = .05$.

c. The p-value $= P(t \geq 1.53)$

Using Table IV, Appendix A, with df = 10,

$.05 < \text{p-value} < .10$

Observing a value of t as large as 1.53 is a rare event if, in fact, $\mu_1 = \mu_2$. Since the probability of this occurring is between .05 and .10, we would conclude that there is some evidence to suggest that $\mu_1 - \mu_2 > 0$.

8.31 $\mu_D = \mu_1 - \mu_2$ = difference in the population means. In testing $H_0: \mu_D = 2$ vs $H_a: \mu_D > 2$, we are testing to see if the difference in the population means $(\mu_1 - \mu_2)$ is greater than 2 or not. Or, we are testing to see if the mean for population 1 is more than 2 units larger than the mean for population 2.

8.33 a.

Person	Difference Before - After
1	83 - 92 = -9
2	60 - 71 = -11
3	55 - 56 = -1
4	99 - 104 = -5
5	77 - 89 = -12

$$\bar{x}_D = \frac{\sum x_D}{n_D} = \frac{-9 + (-11) + (-1) + (-5) + (-12)}{5} = \frac{-38}{5} = -7.6$$

$$s_D^2 = \frac{\sum x_D^2 - \frac{(\sum x_D)^2}{n_D}}{n_D - 1}$$

$$= \frac{(-9)^2 + (-11)^2 + (-1)^2 + (-5)^2 + (-12)^2 - \frac{(-38)^2}{5}}{5 - 1}$$

$$= \frac{372 - 288.8}{4} = 20.8$$

$$s_D = \sqrt{s_D^2} = \sqrt{20.8} = 4.56$$

b. $$\bar{x}_1 = \frac{\sum x_1}{n_1} = \frac{83 + 60 + 55 + 99 + 77}{5} = \frac{374}{5} = 74.8$$

$$\bar{x}_2 = \frac{\sum x_2}{n_2} = \frac{92 + 71 + 56 + 104 + 89}{5} = \frac{412}{5} = 82.4$$

$\bar{x}_1 - \bar{x}_2 = 74.8 - 82.4 = -7.6 = \bar{x}_D$ (part (a))

c. H_0: $\mu_D = 0$ $(\mu_1 = \mu_2)$
H_a: $\mu_D \neq 0$ $(\mu_1 \neq \mu_2)$

The test statistic is $t = \frac{\bar{x}_D - D_0}{s_D/\sqrt{n_D}} = \frac{-7.6 - 0}{4.56/\sqrt{5}} = \frac{-7.6}{2.0393} = -3.73$

The rejection region requires $\alpha/2 = .05/2 = .025$ in each tail of the t distribution with df $= n_D - 1 = 5 - 1 = 4$. From Table IV, Appendix A, $t_{.025} = 2.776$. The rejection region is $t < -2.776$ or $t > 2.776$.

Since the observed value of the test statistic falls in the rejection region ($t = -3.73 < -2.776$), H_0 is rejected. There is sufficient evidence to conclude $\mu_1 \neq \mu_2$ at $\alpha = .05$.

d. p-value = $P(t \leq -3.73) + P(t \geq 3.73)$
$= 2P(t \geq 3.73)$

Using Table IV, Appendix A, with df = 4,

$P(t \geq 3.73)$ is between .01 and .025

Thus, $2P(t \geq 3.73)$ is between .02 and .05

$.02 < \text{p-value} < .05$

Observing a value of t as small as -3.73 or as large as 3.73 is an improbable event, if, in fact, $\mu_D = 0$. Since the probability of this occurring is between .02 and .05, we would conclude that there is some evidence to suggest $\mu_D \neq 0$.

e. For the paired difference test in part (c) to be valid, we need to assume:

1. The relative frequency distribution of the population of differences is normal.
2. The differences are randomly selected from the population of differences.

8.35 Some preliminary calculations are:

Pair	Difference Sample 1 - Sample 2
1	19 - 24 = -5
2	25 - 27 = -2
3	31 - 36 = -5
4	52 - 53 = -1
5	49 - 55 = -6
6	34 - 34 = 0
7	59 - 66 = -7
8	47 - 51 = -4
9	17 - 20 = -3
10	51 - 55 = -4

$$\bar{x}_D = \frac{\sum x_D}{n_D} = \frac{-37}{10} = -3.7$$

$$s_D^2 = \frac{\sum x_D^2 - \frac{(\sum x_D)^2}{n_D}}{n_D - 1} = \frac{181 - \frac{(-37)^2}{10}}{10 - 1} = \frac{44.1}{9} = 4.9$$

$$s_D = \sqrt{s_D^2} = \sqrt{4.9} = 2.2136$$

a. H_0: $\mu_D = 0$
H_a: $\mu_D < 0$ where $\mu_D = \mu_1 - \mu_2$

The test statistic is $t = \dfrac{\bar{x}_D - D_0}{s_D/\sqrt{n_D}} = \dfrac{-3.7 - 0}{2.2136/\sqrt{10}} = \dfrac{-3.7}{.7} = -5.29$

The rejection region requires $\alpha = .05$ in the lower tail of the t distribution with df $= n_D - 1 = 10 - 1 = 9$. From Table IV, Appendix A, $t_{.05} = 1.833$. The rejection region is $t < -1.833$.

Since the observed value of the test statistic falls in the rejection region ($t = -5.29 < -1.833$), H_0 is rejected. There is sufficient evidence to indicate that μ_2 is larger than μ_1 ($\mu_D < 0$) at $\alpha = .05$.

b. For confidence coefficient .90, $\alpha = 1 - .90 = .10$ and $\alpha/2 = .10/2 = .05$. From Table IV, Appendix A, with df $= n_D - 1 = 10 - 1 = 9$, $t_{.05} = 1.833$. The confidence interval is:

$$\bar{x}_D \pm t_{\alpha/2}\, s_D/\sqrt{n_D}$$

$$\Rightarrow -3.7 \pm (1.833)2.2136/\sqrt{10}$$

$$\Rightarrow -3.7 \pm 1.28$$

$$\Rightarrow (-4.98, -2.42)$$

8.37 Some preliminary calculations:

Pair	Difference x - y
1	55 - 44 = 11
2	68 - 55 = 13
3	40 - 25 = 15
4	55 - 56 = -1
5	75 - 62 = 13
6	52 - 38 = 14
7	49 - 31 = 18

$$\bar{x}_D = \frac{\sum x_D}{n_D} = \frac{83}{7} = 11.8571$$

$$s_D^2 = \frac{\sum x_D^2 - \frac{(\sum x_D)^2}{n_D}}{n_D - 1} = \frac{1205 - \frac{83^2}{7}}{7 - 1} = 36.8095$$

$$s_D = \sqrt{s_D^2} = \sqrt{36.8095} = 6.0671$$

a. $H_0: \mu_D = 10$
$H_a: \mu_D \neq 10$ where $\mu_D = (\mu_1 - \mu_2)$

The test statistic is $t = \dfrac{\bar{x}_D - D_0}{s_D/\sqrt{n_D}} = \dfrac{11.8571 - 10}{6.0671/\sqrt{7}} = \dfrac{1.8571}{2.2931} = .81$

The rejection region requires $\alpha/2 = .05/2 = .025$ in each tail of the t distribution with df $= n_D - 1 = 7 - 1 = 6$. From Table IV, Appendix A, $t_{.025} = 2.447$. The rejection region is $t < -2.447$ or $t > 2.447$.

Since the observed value of the test statistic does not fall in the rejection region ($t = .81 \not< -2.447$ and $t = .81 \not> 2.447$), H_0 is not rejected. There is insufficient evidence to conclude $\mu_D \neq 10$ at $\alpha = .05$.

b. p-value $= P(t \leq -.81) + P(t \geq .81)$
$= 2P(t \geq .81)$

Using Table IV, Appendix A, with df = 6,

$P(t \geq .81)$ is greater than .10.

Thus, $2P(t \geq .81)$ is greater than .20.

The probability of observing a value of t as large as .81 or as small as -.81 if, in fact, $\mu_D = 10$ is greater than .20. We would conclude that there is insufficient evidence to suggest $\mu_D \neq 0$.

8.39 Some preliminary calculations are:

Person	Difference Before - After
1	150 - 143 = 7
2	195 - 190 = 5
3	188 - 185 = 3
4	197 - 191 = 6
5	204 - 200 = 4

$$\bar{x}_D = \frac{\sum x_D}{n_D} = \frac{25}{5} = 5$$

$$s_D^2 = \frac{\sum x_D^2 - \frac{(\sum x_D)^2}{n_D}}{n_D - 1} = \frac{135 - \frac{25^2}{5}}{5 - 1} = 2.5$$

$$s_D = \sqrt{s_D^2} = \sqrt{2.5} = 1.5811$$

For confidence coefficient .95, $\alpha = 1 - .95 = .05$ and $\alpha/2 = .05/2 = .025$. From Table IV, Appendix A, with df $= n_D - 1 = 5 - 1 = 4$, $t_{.025} = 2.776$. This is a paired difference experiment since the same individuals are weighed before and after the diet. The 95% confidence interval for $\mu_D = \mu_1 - \mu_2$ is:

$$\bar{x}_D \pm t_{\alpha/2}\, s_D/\sqrt{n_D}$$

$$\Rightarrow 5 \pm 2.776(1.5811)/\sqrt{5}$$

$$\Rightarrow 5 \pm 1.96$$

$$\Rightarrow (3.04, 6.96)$$

In order for the confidence interval above to be valid, we must assume:

1. The relative frequency distribution of the population of differences (Before-After) is normal.
2. The differences are randomly selected from the population of differences.

8.41 Some preliminary calculations are:

Day	Difference Cocoon - Air
1	15.1 - 10.4 = 4.7
2	14.6 - 9.2 = 5.4
3	6.8 - 2.2 = 4.6
4	6.8 - 2.6 = 4.2
5	8.0 - 4.1 = 3.9
6	8.7 - 3.7 = 5.0
7	3.6 - 1.7 = 1.9
8	5.3 - 2.0 = 3.3
9	7.0 - 3.0 = 4.0
10	7.1 - 3.5 = 3.6
11	9.6 - 4.5 = 5.1
12	9.5 - 4.4 = 5.1

$$\bar{x}_D = \frac{\sum x_D}{n_D} = \frac{50.8}{12} = 4.2333$$

$$s_D^2 = \frac{\sum x_D^2 - \frac{(\sum x_D)^2}{n_D}}{n_D - 1} = \frac{225.74 - \frac{50.8^2}{12}}{12 - 1} = \frac{10.6867}{11} = .9715$$

$$s_D = \sqrt{s_D^2} = \sqrt{.9715} = .9857$$

For confidence coefficient .95, $\alpha = 1 - .95 = .05$ and $\alpha/2 = .05/2 = .025$. From Table IV, Appendix A, with $df = n_D - 1 = 12 - 1 = 11$, $t_{.025} = 2.201$. This is a paired difference experiment since the temperature of the air and the caterpillar's body temperature inside the cocoon are taken on the same day.

The 95% confidence interval for $\mu_D = \mu_1 - \mu_2$, where μ_1 = mean body temperature inside the cocoon and μ_2 = mean daily air temperature, is:

$$\bar{x}_D \pm t_{\alpha/2}\, s_D/\sqrt{n_D}$$

$$\Rightarrow 4.23 \pm 2.201(.9857)/\sqrt{12}$$

$$\Rightarrow 4.23 \pm .63$$

$$\Rightarrow (3.60, 4.86)$$

8.43 Some preliminary calculations are:

Car Number	Differences Manufacturer's - Competitor's
1	8.8 - 8.4 = .4
2	10.5 - 10.1 = .4
3	12.5 - 12.0 = .5
4	9.7 - 9.3 = .6
5	9.6 - 9.0 = .6
6	13.2 - 13.0 = .2

$$\bar{x}_D = \frac{\sum x_D}{n_D} = \frac{2.5}{6} = .4167$$

$$s_D^2 = \frac{\sum x_D^2 - \frac{(\sum x_D)^2}{n_D}}{n_D - 1} = \frac{1.13 - \frac{(2.5)^2}{6}}{6 - 1} = \frac{.08833}{5} = .01767$$

$$s_D = \sqrt{s_D^2} = \sqrt{.0177} = .1329$$

a. Let μ_1 = mean strength of the manufacturer's shocks and μ_2 = mean strength of the competitor's shocks. Since we want to determine if there is a difference in the mean strength of the two types of shocks, we test:

$$H_0: \mu_D = 0$$
$$H_a: \mu_D \neq 0 \qquad (\mu_D = \mu_1 - \mu_2)$$

The test statistic is $t = \dfrac{\bar{x}_D - D_0}{s_D/\sqrt{n_D}} = \dfrac{.4167 - 0}{.1329/\sqrt{6}} = \dfrac{.4167}{.05426} = 7.68$

The rejection region requires $\alpha/2 = .05/2 = .025$ in each tail of the t distribution with df $= n_D - 1 = 6 - 1 = 5$. From Table IV, Appendix A, $t_{.025} = 2.571$. The rejection region is $t < -2.571$ or $t > 2.571$.

Since the observed value of the test statistic falls in the rejection region ($t = 7.68 > 2.571$), H_0 is rejected. There is sufficient evidence to conclude there is a difference in the mean strength of the two types of shocks ($\mu_D \neq 0$) at $\alpha = .05$.

b. In order to apply a paired difference analysis to the data, we must assume:

1. The relative frequency distribution of the differences in mean strength of the shocks is normal.
2. The sample of 6 differences are randomly selected from the population of differences.

c. For confidence coefficient .95, $\alpha = 1 - .95 = .05$ and $\alpha/2 = .05/2 = .025$. From Table IV, Appendix A, with df $= n_D - 1 = 6 - 1 = 5$, $t_{.025} = 2.571$. The confidence interval for $\mu_D = \mu_1 - \mu_2$ is:

$$\bar{x}_D \pm t_{\alpha/2}\, s_D/\sqrt{n_D}$$

$$\Rightarrow .42 \pm 2.571\,(.1329)/\sqrt{6}$$

$$\Rightarrow .42 \pm .14$$

$$\Rightarrow (.28,\ .56)$$

We estimate the difference in the average strength of the two types of shocks after 20,000 miles of use falls between .28 and .56. In other words, we estimate that the average strength of the manufacturer's shocks is larger than the average strength of the competitor's shocks by .28 to .56.

8.45 a. Each of six trained female runners were given a liquid 45 minutes prior to running for 85 minutes or until they reached a state of exhaustion. Four liquids were used in the experiment: fructose, glucose, placebo, and water alone. Each of the six runners performed the run for each of the 4 liquids, given in a random order. This experiment does not satisfy independent sampling since the same 6 runners ran with the different liquids taken 45 minutes prior. To be independent sampling, the trained female runners should be randomly divided into different groups and each group administered a different liquid prior to running.

b. Let μ_1 = mean time to exhaustion for runners given the glucose mixture and μ_2 = mean time to exhaustion for runners given the placebo. Since we wish to determine if there is a difference in the mean time to exhaustion between runners using the glucose mixture and those given the placebo, we test:

$H_0: \mu_1 - \mu_2 = 0$
$H_a: \mu_1 - \mu_2 \neq 0$

The test statistic is (for independent samples):

$$t = \frac{(\bar{x}_1 - \bar{x}_2) - D_0}{\sqrt{s_p^2\left(\frac{1}{n_1} + \frac{1}{n_2}\right)}}$$

where $s_p^2 = \frac{s_1^2 + s_2^2}{2} = \frac{20.3^2 + 13.5^2}{2} = 297.17$ (since $n_1 = n_2$)

$$t = \frac{(63.9 - 52.2) - 0}{\sqrt{297.17\left(\frac{1}{6} + \frac{1}{6}\right)}} = \frac{11.7}{9.9527} = 1.18$$

The rejection region requires $\alpha/2 = .05/2 = .025$ in each tail of the t distribution with df $= n_1 + n_2 - 2 = 6 + 6 - 2 = 10$. From Table IV, Appendix A, $t_{.025} = 2.228$ with 10 df. The rejection region is $t < -2.228$ or $t > 2.228$.

Since the observed value of the test statistic does not fall in the rejection region ($t = 1.18 \nless -2.228$ and $t = 1.18 \ngtr 2.228$), H_0 is not rejected. There is insufficient evidence to indicate a difference in the mean time to exhaustion between runners given the glucose mixture and those given the placebo at $\alpha = .05$.

c. The experiment was actually a paired difference experiment. By using the method of part (b), we had a large variance ($s_p^2 = 297.17$) which made it very hard to reject H_0. There were more degrees of freedom but not enough to counteract the large variance. The method used in part (b) should not be used for this particular problem since the assumption of independent samples is violated.

8.47 For confidence coefficient .95, $\alpha = 1 - .95 = .05$ and $\alpha/2 = .05/2 = .025$. From Table III, Appendix A, $z_{.025} = 1.96$.

a. $$n_1 = n_2 = \frac{(z_{\alpha/2})^2(\sigma_1^2 + \sigma_2^2)}{B^2} = \frac{(1.96)^2(13^2 + 14^2)}{2.8^2} = 178.85 \Rightarrow 179$$

b. If the range of each population is 40, we would estimate σ by:

$\sigma \approx 40/4 = 10$

For confidence coefficient .99, $\alpha = 1 - .99 = .01$ and $\alpha/2 = .01/2 = .005$. From Table III, Appendix A, $z_{.005} = 2.58$.

$$n_1 = n_2 = \frac{(z_{\alpha/2})(\sigma_1^2 + \sigma_2^2)}{B^2} = \frac{(2.58)^2(10^2 + 10^2)}{5^2} = 53.25 \Rightarrow 54$$

c. For confidence coefficient .9, $\alpha = 1 - .9 = .1$ and $\alpha/2 = .1/2 = .05$. From Table III, Appendix A, $z_{.05} = 1.645$.

$$n_1 = n_2 = \frac{4(z_{\alpha/2})^2(\sigma_1^2 + \sigma_2^2)}{W^2} = \frac{4(1.645)^2(4.4 + 6.7)}{1.0^2} = 120.15 \Rightarrow 121$$

8.49 If the range of daily incomes for both groups is \$200, we would estimate σ by:

$$\sigma \approx 200/4 = 50$$

For confidence coefficient .9, $\alpha = 1 - .9 = .1$ and $\alpha/2 = .1/2 = .05$. From Table III, Appendix A, $z_{.05} = 1.645$.

$$n_1 = n_2 = \frac{(z_{\sigma/2})^2(\sigma_1^2 + \sigma_2^2)}{B^2} = \frac{(1.645)^2(50^2 + 50^2)}{10^2} = 135.5 \Rightarrow 136$$

8.51 For confidence coefficient .9, $\alpha = 1 - .9 = .1$ and $\alpha/2 = .1/2 = .05$. From Table III, Appendix A, $z_{.05} = 1.645$.

$$n_1 = n_2 = \frac{(z_{\alpha/2})^2(\sigma_1^2 + \sigma_2^2)}{B^2} = \frac{(1.645)^2(5^2 + 5^2)}{1^2} = 135.3 \Rightarrow 136$$

8.53 For confidence coefficient .95, $\alpha = 1 - .95 = .05$ and $\alpha/2 = .05/2 = .025$. From Table III, Appendix A, $z_{.025} = 1.96$.

$$n_1 = n_2 = \frac{4(z_{\alpha/2})^2(\sigma_1^2 + \sigma_2^2)}{W^2}$$

$$= \frac{4(1.96)^2(.05 + .04)}{.2^2}$$

$$= \frac{1.382976}{.04}$$

$$= 34.5744 \approx 35$$

8.55 The Wilcoxon rank sum test is a test of the location (center) of a distribution. The one-tailed test deals specifically with the center of one distribution being shifted in one direction (right or left) from the other distribution. The two-tailed test does not specify a particular direction of shift; we consider the possibility of a shift in either direction.

8.57 a. We first rank all the data:

POPULATION 1		POPULATION 2	
OBSERVATION	RANK	OBSERVATION	RANK
9.0	3	10.1	5
21.1	22	12.0	8
24.8	27	9.2	4
17.2	17	15.8	15
18.9	20	8.8	2
15.6	14	11.1	7
26.9	33	18.2	18
16.5	16	7.0	1
30.1	34	13.6	12
25.4	29	12.5	9
25.6	30	13.5	11
24.6	26	10.3	6
26.0	31	14.2	13
18.7	19	13.2	10
22.0	24	21.5	23
31.1	35		
20.0	21		
25.1	28		
26.1	32		
23.3	25		
	$T_A = 486$		$T_B = 144$

To determine whether the probability distribution for population 2 is shifted to the left of that of population 1, we test:

H_0: The two sampled populations have identical probability distributions
H_a: The probability distribution for population 2 is shifted to the left of that of population 1

The test statistic is

$$z = \frac{T_B - \frac{n_2(n_1 + n_2 + 1)}{2}}{\sqrt{\frac{n_1 n_2(n_1 + n_2 + 1)}{12}}} = \frac{144 - \frac{15(20 + 15 + 1)}{2}}{\sqrt{\frac{20(15)(20 + 15 + 1)}{12}}}$$

$$= \frac{-126}{30} = -4.2$$

The rejection region requires $\alpha = .05$ in the lower tail of the z distribution. From Table III, Appendix A, $z_{.05} = 1.645$. The rejection region is $z < -1.645$.

Since the observed value of the test statistic falls in the rejection region ($z = -4.2 < -1.645$), H_0 is rejected. There is sufficient evidence to indicate the probability distribution for population 2 is shifted to the left of that of population 1 at $\alpha = .05$.

b. The p-value is $P(z \leq -4.2) < .5 - .4990 = .001$ from Table III, Appendix A.

8.59

NEIGHBORHOOD A	RANK	NEIGHBORHOOD B	RANK
.850	11	.911	16
1.060	18	.770	3
.910	15	.815	8
.813	7	.748	2
.737	1	.835	9
.880	13	.800	6
.895	14	.793	4
.844	10	.796	5
.965	17		$T_B = 53$
.875	12		
	$T_A = 118$		

a. H_0: The two sampled neighborhoods have identical probability distributions

H_a: The probability distribution for neighborhood A is shifted to the right or left of neighborhood B

The test statistic is $T_B = 53$.

The null hypothesis will be rejected if $T_B \leq T_L$ or $T_B \geq T_U$ where $\alpha = .05$ (two-tailed), $n_1 = 8$ and $n_2 = 10$. From Table V, Appendix A, $T_L = 54$ and $T_U = 98$.

Reject H_0 if $T_B \leq 54$ or $T_B \geq 98$.

Since $T_B = 53 \leq 54$, we reject H_0 and conclude there is sufficient evidence to indicate neighborhood A is shifted to the right or left of neighborhood B at $\alpha = .05$.

b. The two independent sample t-test is based on the assumptions of:

1) random, independent samples from
2) normally distributed populations with
3) equal variances.

The assumption of normal populations would be necessary to use the t-test.

c. 1) The two samples are random and independent.
2) The two probability distributions from which the samples are drawn are continuous.

8.61

TEST A	RANK	TEST B	RANK
90	12	66	4
71	6	78	8
83	11	50	1
82	10	68	5
75	7	80	9
91	13	60	2
65	3		$T_B = 29$
	$T_A = 62$		

H_0: The probability distributions of scores for tests A and B are identical

H_a: There is a shift in the locations of the probability distributions of scores for tests A and B

The test statistic is $T_B = 29$.

The null hypothesis will be rejected if $T_B \leq T_L$ or $T_B \geq T_U$ where $\alpha = .05$ (two-tailed), $n_1 = 6$ and $n_2 = 7$. From Table V, Appendix A, $T_L = 28$ and $T_U = 56$.

Reject H_0 if $T_B \leq 28$ or $T_B \geq 56$.

Since $T_B = 29 \nleq 28$ and $T_B = 29 \ngeq 56$, do not reject H_0. There is insufficient evidence to indicate a shift in location for tests A and B at $\alpha = .05$.

8.63

DEAF CHILDREN	RANK	HEARING CHILDREN	RANK
2.75	18	1.15	1
3.14	19	1.65	6
3.23	20	1.43	4
2.30	15	1.83	8.5
2.64	17	1.75	7
1.95	10	1.23	2
2.17	13	2.03	12
2.45	16	1.64	5
1.83	8.5	1.96	11
2.23	14	1.37	3
	$T_A = 150.5$		$T_B = 59.5$

H_0: The probability distributions of eye movement rates are identical for deaf and hearing children

H_a: The probability distribution of eye movement rates for deaf children lies to the right of that for hearing children

The test statistic can be either T_A or T_B since the sample sizes are equal. Use $T_A = 150.5$.

The null hypothesis will be rejected if $T_A \geqq T_U$ where $\alpha = .05$ (one-tailed), $n_1 = 10$ and $n_2 = 10$. From Table V, Appendix A, $T_U = 127$.

Reject H_0 if $T_A \geqq 127$.

Since $T_A = 150.5 \geqq 127$, we reject H_0. There is sufficient evidence to indicate that deaf children have greater visual acuity than hearing children at $\alpha = .05$.

8.65 a. The test statistic for this two-tailed test is T, the smaller of the positive and negative rank sums, T_+ and T_-.

The null hypothesis of identical probability distributions will be rejected if $T \leqq T_0$ where T_0 is found in Table VI corresponding to $\alpha = .10$ (two-tailed) and $n = 25$:

Reject H_0 if $T \leqq 101$.

Note: Since $n \geqq 25$, the large sample approximation could also be used.

b. The test statistic is T_-, the negative rank sum.

Since it is necessary to reject the null hypothesis only if the distribution for A is shifted to the right of the distribution for B, small values of T_- would imply rejection of H_0. We will reject H_0 if $T_- \leqq T_0$ where T_0 is found in Table VI corresponding to $\alpha = .05$ (one-tailed) and $n = 41$:

Reject H_0 if $T_- \leqq 303$.

Note: Since $n \geqq 24$, the large sample approximation could also be used.

c. The test statistic is T_+, the positive rank sum.

Since it is necessary to reject the null hypothesis of identical probability distributions only if the distribution of A is shifted to the left of the distribution of B, small values of T_+ would imply rejection of H_0. We will reject H_0 if $T_+ \leqq T_0$ where T_0 is found in Table VI corresponding to $\alpha = .005$ (one-tailed) and $n = 8$:

Reject H_0 if $T_+ \leqq 0$.

8.67 The Wilcoxon signed rank test is a test of the location (center) of a distribution. The one-tailed test deals specifically with the center of one distribution being shifted in one direction (right or left) from the other distribution. The two-tailed test does not specify a particular direction of shift, only that there is a difference in the locations of the two distributions.

8.69

PAIR	SAMPLE FROM POPULATION 1	SAMPLE FROM POPULATION 2	DIFFERENCE	RANK OF ABSOLUTE DIFFERENCE
1	8	7	1	1
2	10	1	9	8
3	6	4	2	2.5
4	10	10	0	(eliminated)
5	7	4	3	4
6	8	3	5	6
7	4	6	-2	2.5
8	9	2	7	7
9	8	4	4	5
			Negative rank sum T_- =	2.5

a. H_0: The probability distributions of the two populations are identical

H_a: The probability distribution of population 1 is shifted to the right of the probability distribution of population 2

Test statistic: $T_- = 2.5$

Reject H_0 if $T_- \leq T_0$ where T_0 is based on $\alpha = .05$ and $n = 8$ (one-tailed):

Reject H_0 if $T_- \leq 6$. (from Table VI, Appendix A).

Conclusion: Reject H_0 at $\alpha = .05$. There is sufficient evidence to conclude that the probability distribution for population 1 is shifted to the right of the probability distribution for population 2.

b. H_0: The probability distributions of the two populations are identical

H_a: The probability distribution of population 1 is shifted either to the right or to the left of the probability distribution of population 2

Test statistic: $T = T_- = 2.5$

Reject H_0 if $T \leq T_0$ where T_0 is based on $\alpha = .05$ and $n = 8$ (two-tailed):

Reject H_0 if $T \leq 4$. (from Table VI, Appendix A).

Conclusion: Reject H_0 at $\alpha = .05$. There is sufficient evidence to conclude that the probability distribution for population 1 is shifted either to the right or to the left of the probability distribution for population 2.

8.71 a. Notice this is an extension of the paired difference experiment where the groups consist of four individuals rather than just one. Thus, we have a randomized block design.

b. H_0: The two sampled populations have identical probability distributions

H_a: The probability distribution for population A (face-to-face) is shifted to the left of that for population B (video teleconferencing)

c.

GROUP	FACE-TO-FACE	VIDEO TELECONFERENCING	DIFFERENCE	RANK OF ABSOLUTE DIFFERENCE
1	65	75	-10	7.5
2	82	80	2	1
3	54	60	-6	6
4	69	65	4	2.5
5	40	55	-15	9
6	85	90	-5	4.5
7	98	98	0	(eliminated)
8	35	40	-5	4.5
9	85	89	-4	2.5
10	70	80	-10	7.5

Negative rank sum $T_- = 41.5$
Positive Rank sum $T_+ = 3.5$

The test statistic is $T_+ = 3.5$.

The null hypothesis will be rejected if $T_+ \leq T_0$ where T_0 corresponds to $\alpha = .05$ (one-tailed) and $n = 9$. From Table VI, Appendix A, $T_0 = 8$.

Reject H_0 if $T_+ \leq 8$.

Since $T_+ = 3.5 \leq 8$, we reject H_0 and conclude there is sufficient evidence to indicate that problem-solving performance of video teleconferencing groups is superior to that of groups that interact face-to-face at $\alpha = .05$.

d. p-value $= P(T_+ \leq 3.5)$ where $n = 9$ and the test is one-tailed. Using Table VI, locate the appropriate column for n, then find the values in that column that include the test statistic (in this case, 6 and 3). Then read the α level corresponding to these values. Thus,

$.01 < \text{p-value} < .025$

8.73

CITY	PERCENTAGE CHANGE IN WHITE STUDENTS			
	1968-1970	1970-1972	DIFFERENCE	RANK OF ABSOLUTE DIFFERENCE
Boston	-3.9	-7.4	3.5	3.5
Philadelphia	-5.7	-2.2	-3.5	3.5
Baltimore	-5.6	-9.3	3.7	5
St. Louis	-9.4	-13.8	4.4	6.5
Chicago	-9.0	-14.7	5.7	8
Detroit	-15.6	-14.0	-1.6	1
Atlanta	-22.3	-34.4	12.1	12
Charlotte	-3.1	-5.6	2.5	2
Jacksonville	-1.8	-11.4	9.6	11
Houston	-9.1	-17.5	8.4	9
San Francisco	-13.5	-22.4	8.9	10
San Diego	-1.1	-5.5	4.4	6.5
				$T_+ = 73.5$
				$T_- = 4.5$

H_0: The percentage change of white students is the same over the two time periods

H_a: The percentage change during 1970-1972 is shifted to the right of the percentage change during 1968-1970

The test statistic is $T_- = 4.5$.

Reject H_0 if $T_- \leq T_0$ where T_0 is based on $n = 12$ and $\alpha = .05$ (one-tailed):

Reject H_0 if $T_- \leq 17$ (from Table VI, Appendix A).

Since $T_- = 4.5 \leq 17$, we reject H_0 and conclude there is sufficient evidence to indicate the shift in white families out of the inner cities was larger in 1970-1972 than for 1968-1970 at $\alpha = .05$.

8.75

TWIN PAIR	SCHOOL A	SCHOOL B	DIFFERENCE	RANK OF ABSOLUTE DIFFERENCE
1	65	69	-4	2
2	72	72	0	(Eliminated)
3	86	74	12	4
4	50	52	-2	1
5	60	47	13	5
6	81	72	9	3
				$T_+ = 12$
				$T_- = 3$

H_0: The achievement test scores are the same for both junior high schools

H_a: The achievement test scores of junior high school A are shifted to the right or left of those for junior high school B

The test statistic is $T_- = 3$.

Reject H_0 if $T \leq T_0$ where T_0 is based on $n = 5$ and $\alpha = .10$ (two-tailed):

Reject H_0 if $T \leq 1$ (from Table VI, Appendix A).

Since $T = 3 \nleq 1$, we fail to reject H_0 and conclude there is no difference in the probability distributions of achievement test scores at the two schools at $\alpha = .10$.

8.77

CHARCOAL CANISTER NO.	PCHD	EERF	DIFFERENCE	RANK OF ABSOLUTE DIFFERENCE
71	1,709.79	1,479.0	230.79	11
58	357.17	257.8	99.37	2
84	1,150.94	1,287.0	-136.06	5
91	1,572.69	1,395.0	177.69	9
44	558.33	416.5	141.83	8
43	4,132.28	3,993.0	139.28	7
79	1,489.86	1,351.0	138.86	6
61	3,017.48	1,813.0	1,204.48	15
85	393.55	187.7	205.85	10
46	880.84	630.4	250.44	12
4	2,996.49	3,707.0	-710.51	14
20	2,367.40	2,791.0	-423.60	13
36	599.84	706.8	-106.96	3
42	538.37	618.5	-80.13	1
55	2,770.23	2,639.0	131.23	4
				$T_+ = 84$
				$T_- = 36$

H_0: The exhalation rate measurements for the two facilities are the same
H_a: The exhalation rate measurements by PCHD are shifted to the right or left of those by EERF

The test statistic is $T_- = 36$.

Reject H_0 if $T \leq T_0$ where T_0 is based on $n = 15$ and $\alpha = .05$ (two-tailed):

Reject H_0 if $T \leq 25$ (from Table VI, Appendix A).

Since $T = 36 \nleq 25$, do not reject H_0. There is insufficient evidence to indicate a difference in the exhalation rate measurements for the two facilities at $\alpha = .05$.

8.79 a. For confidence coefficient .90, $\alpha = .10$ and $\alpha/2 = .05$. From Table III, Appendix A, $z_{.05} = 1.645$. The confidence interval is:

$$(\bar{x}_1 - \bar{x}_2) \pm z_{.05}\sqrt{\frac{s_1^2}{n_1} + \frac{s_2^2}{n_2}}$$

$$\Rightarrow (12.2 - 8.3) \pm 1.645\sqrt{\frac{2.1}{135} + \frac{3.0}{148}}$$

$$\Rightarrow 3.90 \pm .31 \Rightarrow (3.59, 4.21)$$

b. $H_0: \mu_1 - \mu_2 = 0$
$H_a: \mu_1 - \mu_2 \neq 0$

$$\text{Test statistic: } z = \frac{(\bar{x}_1 - \bar{x}_2)}{\sqrt{\frac{s_1^2}{n_1} + \frac{s_2^2}{n_2}}} = \frac{12.2 - 8.3}{\sqrt{\frac{2.1}{135} + \frac{3.0}{148}}} = 20.60$$

The rejection region requires $\alpha/2 = .01/2 = .005$ in each tail of the z distribution. From Table III, Appendix A, $z_{.005} = 2.58$. The rejection region is $z < -2.58$ or $z > 2.58$.

Since the observed value of the test statistic falls in the rejection region ($20.60 > 2.58$), H_0 is rejected. There is sufficient evidence to indicate that $\mu_1 \neq \mu_2$ at $\alpha = .01$.

c. $$n_1 = n_2 = \frac{(z_{\alpha/2})^2(\sigma_1^2 + \sigma_2^2)}{B^2} = \frac{(1.645)^2(2.1 + 3.0)}{.2^2} = 345.02 \Rightarrow 346$$

8.81 a. The 2 samples are randomly selected in an independent manner from the two populations. The sample sizes, n_1 and n_2, are large enough so that $\bar{x}_1$ and $\bar{x}_2$ each have approximately normal sampling distributions and so that s_1^2 and s_2^2 provide good approximations to σ_1^2 and σ_2^2. This will be true if $n_1 \geq 30$ and $n_2 \geq 30$.

b. 1. Both sampled populations have relative frequency distributions that are approximately normal.
   2. The population variances are equal.
   3. The samples are randomly and independently selected from the populations.

c. 1. The relative frequency distribution of the population of differences is normal.
   2. The sample of differences are randomly selected from the population of differences.

8.83 To determine if the students at school B were absent or late less frequently then the students at school A, we test:

$H_0: \mu_1 - \mu_2 = 0$
$H_a: \mu_1 - \mu_2 > 0$

where μ_1 = mean number of late or absent days for school A,
and μ_2 = mean number of late or absent days for school B.

$$\text{Test statistic: } z = \frac{(\bar{x}_1 - \bar{x}_2)}{\sqrt{\frac{s_1^2}{n_1} + \frac{s_2^2}{n_2}}} = \frac{(20.5 - 19.6) - 0}{\sqrt{\frac{26.2}{50} + \frac{24.1}{50}}} = .90$$

The rejection region requires $\alpha = .10$ in the upper tail of the z distribution. From Table III, Appendix A, $z_{.10} = 1.28$. The rejection region is $z > 1.28$.

Since the observed value of the test statistic does not fall in the rejection region ($.90 \not> 1.28$), H_0 is not rejected. There is insufficient evidence to indicate that students from school B were late or absent less frequently than those from school A at $\alpha = .10$.

8.85 Some preliminary calculations are:

Firm	Differences 1989 - 1988
1	5.1 - 6.3 = -1.2
2	104.0 - 116.9 = -12.9
3	6.6 - 8.5 = -1.9
4	17.0 - 14.1 = 2.9
5	1.8 - 2.3 = -.5
6	274.7 - 271.0 = 3.7
7	83.1 - 80.2 = 2.9
8	46.9 - 37.5 = 9.4

$$\bar{x}_D = \frac{\sum x_D}{n_D} = \frac{2.4}{8} = .3$$

$$s_D^2 = \frac{\sum x_D^2 - \frac{(\sum x_D)^2}{n_D}}{n_D - 1} = \frac{290.58 - \frac{2.4^2}{8}}{8 - 1} = \frac{289.86}{7} = 41.4086$$

$$s_D = \sqrt{s_D^2} = \sqrt{41.4086} = 6.4349$$

a. Let μ_1 = mean R & D expenditures for 1989 and μ_2 = mean expenditures for 1988. Since we want to determine if the R & D expenditures have increased, we test:

H_0: $\mu_D = 0$
H_a: $\mu_D > 0$ $\quad (\mu_1 - \mu_2 > 0)$

The test statistic is $t = \frac{\bar{x}_D - D_0}{s_D/\sqrt{n_D}} = \frac{.3 - 0}{6.4349/\sqrt{8}} = .13$

The rejection region requires $\alpha = .10$ in the lower tail of the t distribution with df $= n_D - 1 = 8 - 1 = 7$. From Table IV, Appendix A, $t_{.10} = 1.415$. The rejection region is $t > 1.415$.

Since the observed value of the test statistic does not fall in the rejection region ($t = .13 \not> 1.415$), H_0 is not rejected. There is insufficient evidence to indicate that the mean R & D expenditures have increased from 1988 to 1989 at $\alpha = .10$.

b. A Type I error is rejecting H_0 when H_0 is true. In the context of this problem, a Type I error would occur if we decided that there was an increase in the average R & D expenditures from 1988 to 1989 but there really wasn't an increase.

A Type II error is accepting H_0 when H_0 is false. In the context of this problem, a Type II error would occur if we decided that there was not an increase in the average R & D expenditures from 1988 to 1989 but there really was an increase.

c. In order for the test in part (a) to be valid, we must assume:

1. The relative frequency distribution of the population of differences is normal.
2. The sample of differences are randomly selected from the population of differences.

8.87 Some preliminary calculations are:

TYPE A	RANK	TYPE B	RANK
95	1	110	6
122	10	102	4
102	3	115	8
99	2	112	7
108	5	120	9
	T_A = 21		T_B = 34

To determine if print type A is easier to read, we test:

H_0: The two sampled populations have identical probability distributions

H_a: The probability distribution for print type A is shifted to the left of that for print type B

The test statistic is $T_A = 21$.

The rejection region is $T_A \leq 19$ form Table V, Appendix A, with $n_A = 5$ and $n_B = 5$, and $\alpha = .05$.

Since the observed value of the test statistic does not fall in the rejection region ($T_A = 21 \nleq 19$), H_0 is not rejected. There is insufficient evidence to indicate print type A is easier to read at $\alpha = .05$.

8.89 Some preliminary calculations are:

Patient	Differences Adrenaline - Synthetic
1	3.5 - 3.2 = .3
2	2.6 - 2.8 = -.2
3	3.0 - 3.1 = -.1
4	1.9 - 2.4 = -.5
5	2.9 - 2.9 = 0
6	2.4 - 2.2 = .2
7	2.0 - 2.2 = -.2

$$\bar{x}_D = \frac{\sum x_D}{n_D} = \frac{-.5}{7} = -.0714$$

$$s_D^2 = \frac{\sum x_D^2 - \frac{(\sum x_D)^2}{n_D}}{n_D - 1} = \frac{.47 - \frac{(-.5)^2}{7}}{7 - 1} = \frac{.4343}{6} = .0724$$

$s_D = \sqrt{s_D^2} = \sqrt{.0724} = .2691$

Let μ_1 = mean reduction in eye pressure for adrenaline and μ_2 = mean reduction in eye pressure for the synthetic drug. Since we wish to determine if there is a difference in the mean reductions in eye pressure for the two days, we test:

$H_0\colon \mu_D = 0$
$H_a\colon \mu_D \neq 0 \qquad (\mu_D = \mu_1 - \mu_2)$

The test statistic is $t = \dfrac{\bar{x}_D - D_0}{s_D/\sqrt{n_D}} = \dfrac{-.0714 - 0}{.2691/\sqrt{7}} = \dfrac{-.0714}{.1017} = -.70$

The rejection region requires $\alpha/2 = .10/2 = .05$ in each tail of the t distribution with df $= n_D - 1 = 7 - 1 = 6$. From Table IV, Appendix A, $t_{.05} = 1.943$. The rejection region is $t < -1.943$ or $t > 1.943$.

Since the observed value of the test statistic does not fall in the rejection region ($t = -.70 \nless -1.943$ and $t = .70 \ngtr 1.943$), H_0 is not rejected. The data do not provide sufficient evidence to indicate a difference in the mean reductions in eye pressure for the two drugs at $\alpha = .10$.

8.91 a. Let μ_1 = mean time until spoilage for preservative A and μ_2 = mean time until spoilage for preservative B. Since we wish to determine if there is evidence of a difference in mean time until spoilage begins, we test:

$H_0\colon \mu_1 - \mu_2 = 0$
$H_a\colon \mu_1 - \mu_2 \neq 0$

The test statistic is $t = \dfrac{(\bar{x}_1 - \bar{x}_2) - D_0}{\sqrt{s_p^2\left(\frac{1}{n_1} + \frac{1}{n_2}\right)}}$

where $s_p^2 = \dfrac{s_1^2 + s_2^2}{2} = \dfrac{10.3^2 + 13.4^2}{2} = 142.825$ (since $n_1 = n_2$)

Thus, $t = \dfrac{(106.4 - 96.5) - 0}{\sqrt{142.825\left(\frac{1}{15} + \frac{1}{15}\right)}} = \dfrac{9.9}{4.3639} = 2.27$

The rejection region requires $\alpha/2 = .05/2 = .025$ in each tail of the t distribution with df $= n_1 + n_2 - 2 = 15 + 15 - 2 = 28$. From Table IV, Appendix A, $t_{.025} = 2.048$. The rejection region is $t < -2.048$ or $t > 2.048$.

Since the observed value of the test statistic falls in the rejection region ($t = 2.27 > 2.048$), H_0 is rejected. There is sufficient evidence to indicate a difference in the mean time until spoilage begins between the two preservatives ($\mu_1 - \mu_2 \neq 0$) at $\alpha = .05$.

b. To reduce the variability in the data, the processor could do a paired difference experiment. This could be accomplished by cutting 15 cuts of meat in half and treating one half with preservative A and the other half with preservative B for each cut.

c. For confidence coefficient .95, $\alpha = 1 - .95 = .05$ and $\alpha/2 = .05/2 = .025$. From Table IV, Appendix A, with df $= n_1 + n_2 - 2 = 15 + 15 - 2 = 28$, $t_{.025} = 2.048$. The 95% confidence interval for $(\mu_1 - \mu_2)$ is:

$$(\bar{x}_1 - \bar{x}_2) \pm t_{.025}\sqrt{s_p^2\left(\frac{1}{n_1} + \frac{1}{n_2}\right)}$$

$$\Rightarrow (106.4 - 96.5) \pm 2.048\sqrt{142.825\left(\frac{1}{15} + \frac{1}{15}\right)}$$

$$\Rightarrow 9.9 \pm 8.937 \Rightarrow (.963, 18.837)$$

8.93 Some preliminary calculations are:

Student	Differences Before - After
1	45 - 49 = -4
2	52 - 50 = 2
3	63 - 70 = -7
4	68 - 71 = -3
5	57 - 53 = 4
6	55 - 61 = -6
7	60 - 62 = -2
8	59 - 67 = -8

$$\bar{x}_D = \frac{\sum x_D}{n_D} = \frac{-24}{8} = -3$$

$$s_D^2 = \frac{\sum x_D^2 - \frac{(\sum x_D)^2}{n_D}}{n_D - 1} = \frac{198 - \frac{(-24)^2}{8}}{8 - 1} = 18$$

$$s_D = \sqrt{s_D^2} = \sqrt{18} = 4.2426$$

Let μ_1 = mean test score before the refresher course and μ_2 = mean test score after the refresher course. Since we want to determine if the mean test score has increased, we test:

H_0: $\mu_D = 0$
H_a: $\mu_D < 0$

The test statistic is $t = \dfrac{\bar{x}_D - D_0}{s_D/\sqrt{n_D}} = \dfrac{-3 - 0}{4.2426/\sqrt{8}} = \dfrac{-3}{1.5} = -2.0$

The rejection region requires $\alpha = .05$ in the lower tail of the t distribution with df $= n_D - 1 = 8 - 1 = 7$. From Table IV, Appendix A, $t_{.05} = 1.895$. The rejection region is $t < -1.895$.

Since the observed value of the test statistic falls in the rejection region ($t = -2.0 < -1.895$), H_0 is rejected. There is sufficient evidence to conclude the mean test score has increased ($\mu_D < 0$) at $\alpha = .05$.

8.95 For confidence coefficient .95, $\alpha = 1 - .95 = .05$ and $\alpha/2 = .05/2 = .025$. From Table III, Appendix A, $t_{.025} = 1.96$. Since the range is about 4 at each site, $4\sigma \approx$ Range $\Rightarrow \sigma_1 = \sigma_2 \approx 4/4 = 1$.

$$n_1 = n_2 = \frac{4(z_{\alpha/2})^2(\sigma_1^2 + \sigma_2^2)}{W^2}$$

$$= \frac{4(1.96)^2(1^2 + 1^2)}{.4^2}$$

$$= \frac{30.7328}{.16} = 192.08 \approx 193$$

8.97 Use the Wilcoxon rank sum test.

BEFORE	RANK	AFTER	RANK
26	7.5	28	11.5
25	6	30	15
27	9.5	27	9.5
26	7.5	31	16
24	5	29	13.5
20	1	23	4
22	3	28	11.5
21	2	29	13.5
T_{Before} =	41.5	T_{After} =	94.5

H_0: The probability distributions of number of patients treated per day are identical before and after the group practice

H_a: The probability distribution of number of patients treated per day before joining the group is shifted to the left of that after joining the group practice

Test statistic: $T = T_{After} = 94.5$

The rejection region is $T \geq T_U$, where $T_U = 84$ is the upper value given by Table V in Appendix A for $\alpha = .05$ one-tailed, with $n_1 = 8$, $n_2 = 8$. The rejection region is $T \geq 84$.

Since the observed value of the test statistic falls in the rejection region ($94.5 > 84$), H_0 is rejected. There is sufficient evidence to conclude that the doctor tends to treat more patients per day now than before at $\alpha = .05$.

CHAPTER

9

COMPARING POPULATION PROPORTIONS

9.1 A binomial experiment has the following five characteristics:

1. The experiment consists of n identical trials.
2. There are only two possible outcomes on each trial (Success/Failure).
3. The probability of S on any trial, p, is a constant.
4. The trials are independent.
5. The random variable is the number of S's in n trials.

9.3 a. The rejection region requires $\alpha = .01$ in the lower tail of the z distribution. From Table III, Appendix A, $z_{.01} = 2.33$. The rejection region is $z < -2.33$.

b. The rejection region requires $\alpha = .025$ in the lower tail of the z distribution. From Table III, Appendix A, $z_{.025} = 1.96$. The rejection region is $z < -1.96$.

c. The rejection region requires $\alpha = .05$ in the lower tail of the z distribution. From Table III, Appendix A, $z_{.05} = 1.645$. The rejection region is $z < -1.645$.

d. The rejection region requires $\alpha = .10$ in the lower tail of the z distribution. From Table III, Appendix A, $z_{.10} = 1.28$. The rejection region is $z < -1.28$.

9.5 For confidence coefficient .95, $\alpha = 1 - .95 = .05$ and $\alpha/2 = .05/2 = .025$. From Table III, Appendix A, $z_{.025} = 1.96$. The 95% confidence interval for $p_1 - p_2$ is approximately:

a. $(\hat{p}_1 - \hat{p}_2) \pm z_{\alpha/2}\sqrt{\dfrac{\hat{p}_1\hat{q}_1}{n_1} + \dfrac{\hat{p}_2\hat{q}_2}{n_2}}$

$\Rightarrow (.65 - .58) \pm 1.96\sqrt{\dfrac{.65(1 - .65)}{400} + \dfrac{.58(1 - .58)}{400}}$

$\Rightarrow .07 \pm .067$

$\Rightarrow (.003, .137)$

b. $(\hat{p}_1 - \hat{p}_2) \pm z_{\alpha/2}\sqrt{\frac{\hat{p}_1\hat{q}_1}{n_1} + \frac{\hat{p}_2\hat{q}_2}{n_2}}$

$\Rightarrow (.31 - .25) \pm 1.96\sqrt{\frac{.31(1 - .31)}{180} + \frac{.25(1 - .25)}{250}}$

$\Rightarrow .06 \pm .086$

$\Rightarrow (-.026, .146)$

c. $(\hat{p}_1 - \hat{p}_2) \pm z_{\alpha/2}\sqrt{\frac{\hat{p}_1\hat{q}_1}{n_1} + \frac{\hat{p}_2\hat{q}_2}{n_2}}$

$\Rightarrow (.46 - .61) \pm 1.96\sqrt{\frac{.46(1 - .46)}{100} + \frac{.61(1 - .61)}{120}}$

$\Rightarrow -.15 \pm .131$

$\Rightarrow (-.281, -.019)$

9.7 $H_0: (p_1 - p_2) = .1$
$H_a: (p_1 - p_2) > .1$

Since D_0 is not equal to 0,

the test statistic is $z = \frac{(\hat{p}_1 - \hat{p}_2) - D_0}{\sqrt{\frac{p_1q_1}{n_1} + \frac{p_2q_2}{n_2}}} \approx \frac{(\hat{p}_1 - \hat{p}_2) - D_0}{\sqrt{\frac{\hat{p}_1\hat{q}_1}{n_1} + \frac{\hat{p}_2\hat{q}_2}{n_2}}}$

$$= \frac{(.4 - .2) - .1}{\sqrt{\frac{.4(1 - .4)}{50} + \frac{.2(1 - .2)}{60}}} = \frac{.1}{.0864} = 1.16$$

The rejection region requires $\alpha = .05$ in the upper tail of the z distribution. From Table III, Appendix A, $z_\alpha = 1.645$. The rejection region is $z > 1.645$.

Since the observed value of the test statistic does not fall in the rejection region ($z = 1.16 \not> 1.645$), H_0 is not rejected. There is insufficient evidence to show $(p_1 - p_2) > .1$ at $\alpha = .05$.

9.9 a. Let p_1 = proportion of dead insects in the room sprayed with type A and p_2 = proportion of dead insects in the room sprayed with type B. Since we wish to determine if spray A is more effective than spray B, we test:

$H_0: (p_1 - p_2) = 0$
$H_a: (p_1 - p_2) > 0$

The test statistic is $z = \frac{(\hat{p}_1 - \hat{p}_2)}{\sqrt{\frac{p_1q_1}{n_1} + \frac{p_2q_2}{n_2}}} \approx \frac{(\hat{p}_1 - \hat{p}_2)}{\sqrt{\hat{p}\hat{q}(\frac{1}{n_1} + \frac{1}{n_2})}}$

$$\text{where } \hat{p} = \frac{x_1 + x_2}{n_1 + n_2} = \frac{120 + 80}{200 + 200} = \frac{200}{400} = .5$$

$$\hat{p}_1 = \frac{120}{200} = .6, \quad \hat{p}_2 = \frac{80}{200} = .4$$

$$z = \frac{.6 - .4}{\sqrt{.5(1 - .5)\left(\frac{1}{200} + \frac{1}{200}\right)}} = \frac{.2}{.05} = 4$$

The rejection region requires $\alpha = .05$ in the upper tail of the z distribution. From Table III, Appendix A, $z_{.05} = 1.645$. The rejection region is $z > 1.645$.

Since the observed value of the test statistic falls in the rejection region ($z = 4 > 1.645$), H_0 is rejected. There is sufficient evidence to indicate spray A is more effective than spray B in controlling the insects ($p_1 - p_2 > 0$) at $\alpha = .05$.

b. For confidence coefficient .90, $\alpha = 1 - .90 = .10$ and $\alpha/2 = .10/2 = .05$. From Table III, Appendix A, $z_{.05} = 1.645$. The 90% confidence interval for $(p_1 - p_2)$ is:

$$(\hat{p}_1 - \hat{p}_2) \pm z_{\alpha/2} \sqrt{\frac{p_1 q_1}{n_1} + \frac{p_2 q_2}{n_2}}$$

$$\approx (\hat{p}_1 - \hat{p}_2) \pm z_{\alpha/2} \sqrt{\frac{\hat{p}_1 \hat{q}_1}{n_1} + \frac{\hat{p}_2 \hat{q}_2}{n_2}}$$

$$\Rightarrow (.6 - .4) \pm 1.645\sqrt{\frac{.6(1 - .6)}{200} + \frac{.4(1 - .4)}{200}}$$

$$\Rightarrow .2 \pm .08 \Rightarrow (.12, .28)$$

We estimate the difference in the rates of kill for the two sprays to fall between .12 and .28. In other words, we estimate that p_1, the rate of kill for spray A, is larger than p_2, the rate of kill for spray B by .12 to .28.

9.11 For confidence coefficient .95, $\alpha = 1 - .95 = .05$ and $\alpha/2 = .05/2 = .025$. From Table III, Appendix A, $z_{.025} = 1.96$. Let p_1 = proportion of Fortune 500 CEO's under 45 and p_2 = proportion of entrepreneurs under 45. The 95% confidence interval for $(p_1 - p_2)$ is:

$$(\hat{p}_1 - \hat{p}_2) \pm z_{\alpha/2} \sqrt{\frac{p_1 q_1}{n_1} + \frac{p_2 q_2}{n_2}}$$

$$\approx (\hat{p}_1 - \hat{p}_2) \pm z_{\alpha/2} \sqrt{\frac{\hat{p}_1 \hat{q}_1}{n_1} + \frac{\hat{p}_2 \hat{q}_2}{n_2}}$$

where $\hat{p}_1 = \frac{19}{207} = .092$ and $\hat{p}_2 = \frac{96}{153} = .627$

$$\Rightarrow (.092 - .627) \pm 1.96\sqrt{\frac{.092(1 - .092)}{200} + \frac{.627(1 - .627)}{153}}$$

$\Rightarrow -.535 \pm .086$

$\Rightarrow (-.621, -.449)$

9.13 a. Let p_1 = proportion of the American public that claim to read a book occasionally in 1978 and p_2 = proportion that claim to read a book occasionally in 1983. Since we wish to determine if there is a change in the proportion of people who claim to read a book occasionally from 1978 to 1983, we test:

$H_0: p_1 - p_2 = 0$
$H_a: p_1 - p_2 \neq 0$

The test statistic is $z = \frac{(\hat{p}_1 - \hat{p}_2)}{\sqrt{\frac{p_1 q_1}{n_1} + \frac{p_2 q_2}{n_2}}} \approx \frac{(\hat{p}_1 - \hat{p}_2)}{\sqrt{\hat{p}\hat{q}\left(\frac{1}{n_1} + \frac{1}{n_2}\right)}}$

where $\hat{p}_1 = .55$, $\hat{p}_2 = .56$, $\hat{p} = \frac{\hat{p}_1 + \hat{p}_2}{2} = \frac{.55 + .56}{2} = .555$

(since $n_1 = n_2$)

$$z = \frac{.55 - .56}{\sqrt{.555(1 - .555)\left(\frac{1}{1961} + \frac{1}{1961}\right)}} = \frac{-.01}{.0159} = -.63$$

The rejection region requires $\alpha/2 = .01/2 = .005$ in each tail of the z distribution. From Table III, Appendix A, $z_{.005} = 2.58$. The rejection region is $z < -2.58$ or $z > 2.58$.

Since the observed value of the test statistic does not fall in the rejection region ($z = -.63 \not< -2.58$ and $z = -.63 \not> 2.58$), H_0 is not rejected. There is insufficient evidence to indicate a change from 1978 to 1983 in the proportion of people who claim to occasionally read a book at $\alpha = .01$.

b. For confidence coefficient .99, $\alpha = 1 - .99 = .01$ and $\alpha/2 = .01/2 = .005$. From Table III, Appendix A, $z_{.005} = 2.58$. The 99% confidence interval for $(p_1 - p_2)$ is approximately:

$$(\hat{p}_1 - \hat{p}_2) \pm z_{\alpha/2}\sqrt{\frac{\hat{p}_1\hat{q}_1}{n_1} + \frac{\hat{p}_2\hat{q}_2}{n_2}}$$

$$\Rightarrow (.18 - .35) \pm 2.58\sqrt{\frac{.18(1 - .18)}{1961} + \frac{.35(1 - .35)}{1961}}$$

$\Rightarrow -.17 \pm .036$

$\Rightarrow (-.206, -.134)$

Thus, we estimate the difference in the proportion of heavy readers from 1978 to 1983 is between -.206 to -.134. In other words, we estimate that p_2, the proportion of heavy readers in 1983 is larger than p_1, the proportion of heavy readers in 1978 by .134 to .206.

9.15 a. Let p_1 = proportion that went blind in 1 year with laser beam treatment and p_2 = proportion that went blind in 1 year in the control group. Since we wish to determine if the laser beam treatment was effective in reducing the probability of going blind, we test:

$$H_0: p_1 - p_2 = 0$$
$$H_a: p_1 - p_2 < 0$$

The test statistic is $z = \dfrac{(\hat{p}_1 - \hat{p}_2)}{\sqrt{\dfrac{p_1q_1}{n_1} + \dfrac{p_2q_2}{n_2}}} \approx \dfrac{(\hat{p}_1 - \hat{p}_2)}{\sqrt{\hat{p}\hat{q}\left(\dfrac{1}{n_1} + \dfrac{1}{n_2}\right)}}$

where $\hat{p}_1 = .14$, $\hat{p}_2 = .42$, $\hat{p} = \dfrac{\hat{p}_1 + \hat{p}_2}{2} = \dfrac{.14 + .42}{2} = .28$ (since $n_1 = n_2$)

$$z \approx \frac{.14 - .42}{\sqrt{.28(1 - .28)\left(\frac{1}{224} + \frac{1}{224}\right)}} = \frac{-.28}{.0424} = -6.60$$

The rejection region requires $\alpha = .05$ in the lower tail of the z distribution. From Table III, Appendix A, $z_{.05} = 1.645$. The rejection region is $z < -1.645$.

Since the observed value of the test statistic falls in the rejection region ($z = -6.60 < -1.645$), H_0 is rejected. There is sufficient evidence to indicate that the laser beam treatment was effective in reducing the probability that a patient will go blind after 1 year ($p_1 - p_2 < 0$) at $\alpha = .05$.

b. p-value = $P(z \leq -6.60) \approx .5 - .5 = 0$ (from Table III, Appendix A)

Observing a value of z as small as -6.60 is an improbable event if, in fact, $p_1 = p_2$. Since the probability of this occurring is approximately zero, we would conclude that there is strong evidence to suggest $p_1 - p_2 < 0$.

c. For confidence coefficient .95, $\alpha = 1 - .95 = .05$ and $\alpha/2 = .05/2 = .025$. From Table III, Appendix A, $z_{.025} = 1.96$. The 95% confidence interval for $(p_1 - p_2)$ is approximately:

$$(\hat{p}_1 - \hat{p}_2) \pm z_{\alpha/2}\sqrt{\frac{\hat{p}_1\hat{q}_1}{n_1} + \frac{\hat{p}_2\hat{q}_2}{n_2}}$$

$$\Rightarrow (.14 - .42) \pm 1.96\sqrt{\frac{.14(1 - .14)}{224} + \frac{.42(1 - .42)}{224}}$$

$$\Rightarrow -.28 \pm .079$$

$$\Rightarrow (-.359, -.201)$$

9.17 a. Let p_1 = proportion of miners with breathing irregularities and p_2 = proportion of Duluth men with breathing irregularities. Since we wish to determine if there are a higher proportion of breathing irregularities among the miners, we test:

$H_0: p_1 - p_2 = 0$
$H_a: p_1 - p_2 > 0$

The test statistic is $z = \dfrac{(\hat{p}_1 - \hat{p}_2)}{\sqrt{\dfrac{p_1q_1}{n_1} + \dfrac{p_2q_2}{n_2}}} \approx \dfrac{\hat{p}_1 - \hat{p}_2}{\sqrt{\hat{p}\hat{q}\left(\dfrac{1}{n_1} + \dfrac{1}{n_2}\right)}}$

where $\hat{p}_1 = \dfrac{x_1}{n_1} = \dfrac{61}{307} = .1987$, $\hat{p}_2 = \dfrac{x_2}{n_2} = \dfrac{5}{35} = .1429$, and

$$\hat{p} = \frac{x_1 + x_2}{n_1 + n_2} = \frac{61 + 5}{307 + 35} = \frac{66}{342} = .193$$

$$z = \frac{.1987 - .1429}{\sqrt{.193(1 - .193)\left(\frac{1}{307} + \frac{1}{35}\right)}} = \frac{.0558}{.0704} = .79$$

The rejection region requires $\alpha = .05$ in the upper tail of the z distribution. From Table III, Appendix A, $z_{.05} = 1.645$. The rejection region is $z > 1.645$.

Since the observed value of the test statistic does not fall in the rejection region ($z = .79 \not> 1.645$), H_0 is not rejected. There is insufficient evidence to indicate that a higher proportion of breathing irregularities exists among those exposed to taconite dust at $\alpha = .05$.

b. p-value $= P(z \geq .79) = .5 - .2852 = .2148$ (from Table III, Appendix A).

Since the probability of observing a value of z this large if, in fact, $p_1 = p_2$, is .2148, we would conclude that there is insufficient evidence to conclude $p_1 - p_2 > 0$.

9.19 Let p_1 = proportion who drank coffee in 1984 and p_2 = proportion who drank coffee in 1962. For confidence coefficient .95, $\alpha = 1 - .95 = .05$ and $\alpha/2 = .05/2 = .025$. From Table III, Appendix A, $z_{.025} = 1.96$. The 95% confidence interval for $(p_1 - p_2)$ is approximately:

$$(\hat{p}_1 - \hat{p}_2) \pm z_{\alpha/2}\sqrt{\frac{\hat{p}_1\hat{q}_1}{n_1} + \frac{\hat{p}_2\hat{q}_2}{n_2}}$$

$$\Rightarrow (.552 - .747) \pm 1.96\sqrt{\frac{.552(1 - .552)}{1000} + \frac{.747(1 - .747)}{1000}}$$

$$\Rightarrow -.195 \pm .041$$

$$\Rightarrow (-.236, -.154)$$

9.21 a. $P(\hat{p}_1 - \hat{p}_2 \leq -.15) + P(\hat{p}_1 - \hat{p}_2 \geq .15)$

$$= P\left[z \leq \frac{-.15}{\sqrt{.845(.155)\left(\frac{1}{471} + \frac{1}{471}\right)}}\right] + P\left(z \geq \frac{.15}{\sqrt{.845(.155)\left(\frac{1}{471} + \frac{1}{471}\right)}}\right)$$

$$= P(z \leq -6.36) + P(z \geq 6.36)$$

$$\approx (.5 - .5) + (.5 - .5) \quad \text{(From Table IV, Appendix A)}$$

$$= 0$$

b. Let p_1 = proportion of survivals for those that took the placebo within 4 months of their heart attack and p_2 = proportion of survivals for those that took the placebo between 1 and 7-1/2 years after their heart attack. Since we wish to determine if the proportion of survival is equal for the 2 groups,

H_0: $p_1 - p_2 = 0$
H_a: $p_1 - p_2 \neq 0$

The test statistic is $z = \dfrac{(\hat{p}_1 - \hat{p}_2)}{\sqrt{\frac{\hat{p}_1\hat{q}_1}{n_1} + \frac{\hat{p}_2\hat{q}_2}{n_2}}} \approx \dfrac{(\hat{p}_1 - \hat{p}_2)}{\sqrt{\hat{p}\hat{q}\left(\frac{1}{n_1} + \frac{1}{n_2}\right)}}$

where $\hat{p}_1 = .77$, $\hat{p}_2 = .92$, $\hat{p} = \dfrac{\hat{p}_1 + \hat{p}_2}{2} = \dfrac{.77 + .92}{2} = .845$ (since $n_1 = n_2$)

$$z \approx \frac{.77 - .92}{\sqrt{.845(1 - .845)\left(\frac{1}{471} + \frac{1}{471}\right)}} = \frac{-.15}{.0236} = -6.36$$

p-value $= P(z \leq -6.36) + P(z \geq 6.36)$

$= 2P(z \geq 6.36)$

$\approx 2(.5 - .5) = 0$ (From Table III, Appendix A)

Observing a value of z as small as -6.36 or as large as 6.36 is an improbable event if, in fact, $p_1 = p_2$. Since the probability of this occurring is approximately zero, we would conclude that there is strong evidence to suggest $p_1 - p_2 \neq 0$. Therefore, we would reject H_0.

9.23 For confidence coefficient .90, $\alpha = 1 - .90 = .10$ and $\alpha/2 = .10/2 = .05$. From Table III, Appendix A, $z_{.05} = 1.645$. Since we have no prior information about the proportions, we use $p_1 = p_2 = .5$ to get a conservative estimate.

$$n_1 = n_2 = \frac{4(z_{\alpha/2})^2(p_1q_1 + p_2q_2)}{W^2}$$

$$= \frac{4(1.645)^2(.5(1 - .5) + .5(1 - .5))}{.06^2}$$

$$= \frac{5.41205}{.0036}$$

$$= 1503.35 \approx 1504$$

9.25 a. For confidence coefficient .80, $\alpha = 1 - .80 = .20$ and $\alpha/2 = .20/2 = .10$. From Table III, Appendix A, $z_{.10} = 1.28$. Since we have no prior information about the proportions, we use $p_1 = p_2 = .5$ to get a conservative estimate.

$$n_1 = n_2 = \frac{4(z_{\alpha/2})^2(p_1q_1 + p_2q_2)}{W^2}$$

$$= \frac{4(1.28)^2(.5(1 - .5) + .5(1 - .5))}{.06^2}$$

$$= \frac{3.2768}{.0036}$$

$$= 910.22 \approx 911$$

b. For confidence coefficient .90, $\alpha = 1 - .90 = .10$ and $\alpha/2 = .10/2 = .05$. From Table III, Appendix A, $z_{.05} = 1.645$. Using the formula for the sample size needed to estimate a proportion from Chapter 7,

$$n = \frac{(z_{\alpha/2})^2pq}{B^2}$$

$$= \frac{1.645^2(.5(1 - .5))}{.02^2}$$

$$= \frac{.6765}{.0004}$$

$$= 1691.27 \approx 1692$$

No, the sample size from part (a) is not large enough.

9.27 For confidence coefficient .90, $\alpha = 1 - .90 = .10$ and $\alpha/2 = .10/2 = .05$. From Table III, Appendix A, $z_{.05} = 1.645$. Since we want $n_1 = 2n_2$,

$$z_{\alpha/2}\sqrt{\frac{p_1q_1}{n_1} + \frac{p_2q_2}{n_2}} = B$$

$$1.645\sqrt{\frac{.2(1 - .2)}{n_1} + \frac{.2(1 - .2)}{n_2}} = .05$$

$$1.645\sqrt{\frac{.2(.8)}{2n_2} + \frac{.2(.8)}{n_2}} = .05$$

$$1.645\sqrt{\frac{.16 + .32}{2n_2}} = .05$$

$$1.645\sqrt{\frac{.24}{n_2}} = .05$$

$$\sqrt{\frac{.24}{n_2}} = .030395$$

$$\frac{.24}{n_2} = .00092386$$

$$n_2 = 259.78 \approx 260$$

Thus, $n_1 = 2n_2 = 2(260) = 520$ and $n_2 = 260$.

9.29 a. The rejection region requires $\alpha = .10$ in the upper tail of the χ^2 distribution with df = k - 1 = 3 - 1 = 2. From Table VII, Appendix A, $\chi^2_{.10} = 4.60517$. The rejection region is $\chi^2 > 4.60517$.

b. The rejection region requires $\alpha = .01$ in the upper tail of the χ^2 distribution with df = k - 1 = 5 - 1 = 4. From Table VII, Appendix A, $\chi^2_{.01} = 13.2767$. The rejection region is $\chi^2 > 13.2767$.

c. The rejection region requires $\alpha = .05$ in the upper tail of the χ^2 distribution with df = k - 1 = 4 - 1 = 3. From Table VII, Appendix A, $\chi^2_{.05} = 7.81473$. The rejection region is $\chi^2 > 7.81473$.

9.31 The sample size n will be large enough so that for every cell the expected cell count $E(n_i)$ will be equal to 5 or more.

9.33 Some preliminary calculations are:

If the probabilities are the same, $p_{1,0} = p_{2,0} = p_{3,0} = p_{4,0} = .25$

$E(n_1) = np_{1,0} = 206(.25) = 51.5$
$E(n_2) = E(n_3) = E(n_4) = 206(.25) = 51.5$

a. To determine if the multinomial probabilities differ, we test:

H_0: $p_1 = p_2 = p_3 = p_4 = .25$
H_a: At least one of the probabilities differs from .25

The test statistic is $\chi^2 = \sum \frac{[n_i - E(n_i)]^2}{E(n_i)}$

$$= \frac{(45 - 51.5)^2}{51.5} + \frac{(54 - 51.5)^2}{51.5} + \frac{(60 - 51.5)^2}{51.5} + \frac{(47 - 51.5)^2}{51.5}$$

$$= 2.738$$

The rejection region requires $\alpha = .05$ in the upper tail of the χ^2 distribution with df = k - 1 = 4 - 1 = 3. From Table VII, Appendix A, $\chi^2_{.05} = 7.81473$. The rejection region is $\chi^2 > 7.81473$.

Since the observed value of the test statistic does not fall in the rejection region ($\chi^2 = 2.7379 \not> 7.81473$), H_0 is not rejected. There is insufficient evidence to indicate the multinomial probabilities differ at $\alpha = .05$.

b. The Type I error is concluding the multinomial probabilities differ when, in fact, they do not.

The Type II error is concluding the multinomial probabilities are equal, when, in fact, they are not.

9.35 a. To determine if a difference in the preference of entrepreneurs for the cars, we test:

H_0: $p_1 = p_2 = p_3 = 1/3$
H_a: At least one of the multinomial probabilities does not equal 1/3

Where p_1 = proportion who prefer U.S. cars
p_2 = proportion who prefer European cars
p_3 = proportion who prefer Japanese cars

$$E(n_1) = np_{1,0} = 100(\frac{1}{3}) = 33\frac{1}{3}$$

$$E(n_2) = np_{2,0} = 100(\frac{1}{3}) = 33\frac{1}{3}$$

$$E(n_3) = np_{3,0} = 100(\frac{1}{3}) = 33\frac{1}{3}$$

The test statistic is $\chi^2 = \sum \frac{[n_i - E(n_i)]^2}{E(n_i)} = \frac{(45 - 33\frac{1}{3})^2}{33\frac{1}{3}}$

$$+ \frac{(46 - 33\frac{1}{3})^2}{33\frac{1}{3}} + \frac{(9 - 33\frac{1}{3})^2}{33\frac{1}{3}} = 26.66$$

The rejection region requires $\alpha = .05$ in the upper tail of the χ^2 distribution with df = k - 1 = 3 - 1 = 2. From Table VII, Appendix A, $\chi^2_{.05} = 5.99147$. The rejection region is $\chi^2 > 5.99147$.

Since the observed value of the test statistic falls in the rejection region ($\chi^2 = 26.66 > 5.99147$), H_0 is rejected. There is sufficient evidence to indicate a difference in the preference of entrepreneurs for the cars at $\alpha = .05$.

b. To determine if there is a difference in the preference of entrepreneurs for domestic versus foreign cars, we test:

H_0: $p_1 = p_2 = .5$
H_a: At least one probability differs from .5.

Where p_1 = proportion who prefer U.S. cars
p_2 = proportion who prefer foreign cars

$E(n_1) = np_{1,0} = 100(.5) = 50$
$E(n_2) = np_{2,0} = 100(.5) = 50$

The test statistic is $\chi^2 = \sum\frac{[n_i - E(n_i)]^2}{E(n_i)} = \frac{(45 - 50)^2}{50} + \frac{(55 - 50)^2}{50} = 1$

The rejection region requires $\alpha = .05$ in the upper tail of the χ^2 distribution with df = k - 1 = 2 - 1 = 1. From Table VII, Appendix A, $\chi^2_{.05} = 3.84146$. The rejection region is $\chi^2 > 3.84146$.

Since the observed value of the test statistic does not fall in the rejection region ($\chi^2 = 1 \not> 3.84146$), H_0 is not rejected. There is insufficient evidence to indicate a difference in the preference of entrepreneurs for domestic and foreign cars at $\alpha = .05$.

c. We must assume the sample size n is large enough so that the expected cell count, $E(n_i)$, will be equal to 5 or more for every cell.

d. For confidence coefficient .90, $\alpha = 1 - .90 = .10$ and $\alpha/2 = .10/2 = .05$. From Table III, Appendix A, $z_{.05} = 1.645$. The 90% confidence interval is

$$\hat{p} \pm z_{\alpha/2}\sqrt{\frac{\hat{p}\hat{q}}{n}} \quad \text{where } \hat{p} = \frac{46 + 9}{100} = .55$$

$$\Rightarrow .55 \pm 1.645\sqrt{\frac{.55(.45)}{100}} \Rightarrow .55 \pm .082 \Rightarrow (.468, .632)$$

9.37 Some preliminary calculations are:

If the strategies are equally preferred, $p_{1,0} = p_{2,0} = p_{3,0} = p_{4,0} = p_{5,0} = .2$

$$E(n_1) = E(n_2) = E(n_3) = E(n_4) = E(n_5) = np = 100(.2) = 20$$

To determine if there is a preference for one or more of the strategies, we test:

H_0: $p_1 = p_2 = p_3 = p_4 = p_5 = .2$
H_a: At least one probability differs from .2

The test statistic is $\chi^2 = \sum \frac{[n_i - E(n_i)]^2}{E(n_i)}$

$$= \frac{(17 - 20)^2}{20} + \frac{(27 - 20)^2}{20} + \frac{(22 - 20)^2}{20} + \frac{(15 - 20)^2}{20} + \frac{(19 - 20)^2}{20} = 4.4$$

The rejection region requires $\alpha = .05$ in the upper tail of the χ^2 distribution with df = k - 1 = 5 - 1 = 4. From Table VII, Appendix A, $\chi^2_{.05} = 9.48773$. The rejection region is $\chi^2 > 9.48773$.

Since the observed value of the test statistic does not fall in the rejection region ($\chi^2 = 4.4 \not> 9.48773$), H_0 is not rejected. There is insufficient evidence to indicate one of the strategies is preferred over any of the others at $\alpha = .05$.

9.39 Some preliminary calculations are:

$E(n_1) = np_1 = 144(.25) = 36$
$E(n_2) = np_2 = 144(.5) = 72$
$E(n_3) = np_3 = 144(.25) = 36$

To determine if there is evidence to contradict the geneticist's theory, we test:

H_0: $p_1 = .25$, $p_2 = .50$, and $p_3 = .25$
H_a: At least one of the probabilities differs from its hypothesized value

The test statistic is $\chi^2 = \sum \frac{[n_i - E(n_i)]^2}{E(n_i)}$

$$= \frac{(30 - 36)^2}{36} + \frac{(78 - 72)^2}{72} + \frac{(36 - 36)^2}{36} = 1.5$$

The rejection region requires $\alpha = .05$ in the upper tail of the χ^2 distribution with df = k - 1 = 3 - 1 = 2. From Table VII, Appendix A, $\chi^2_{.05} = 5.991947$. The rejection region is $\chi^2 > 5.99147$.

Since the observed value of the test statistic does not fall in the rejection region ($\chi^2 = 1.5 \not> 5.99147$), H_0 is not rejected. There is insufficient evidence to contradict the geneticist's theory at $\alpha = .05$.

9.41 a. Some preliminary calculations are:

$E(n_1) = np_{1,0} = 300(.9) = 270$
$E(n_2) = np_{2,0} = 300(.04) = 12$
$E(n_3) = np_{3,0} = 300(.03) = 9$
$E(n_4) = np_{4,0} = 300(.02) = 6$
$E(n_5) = np_{5,0} = 300(.005) = 1.5$
$E(n_6) = np_{6,0} = 300(.005) = 1.5$

To determine if the proportions of printed invoices in the six error categories differ from the proportions using the previous format, we test:

H_0: $p_1 = .90$, $p_2 = .04$, $p_3 = .03$, $p_4 = .02$, $p_5 = .005$, and $p_6 = .005$
H_a: At least one of the probabilities differs from its hypothesized value

The test statistic is $\chi^2 = \sum \frac{[n_i - E(n_i)]^2}{E(n_i)} = \frac{(150 - 270)^2}{270}$

$+ \frac{(120 - 12)^2}{12} + \frac{(15 - 9)^2}{9} + \frac{(7 - 6)^2}{6} + \frac{(4 - 1.5)^2}{1.5}$

$+ \frac{(4 - 1.5)^2}{1.5} = 1037.833$

The rejection region requires $\alpha = .05$ in the upper tail of the χ^2 distribution with df $= k - 1 = 6 - 1 = 5$. From Table VII, Appendix A, $\chi^2_{.05} = 11.0705$. The rejection region is $\chi^2 > 11.0705$.

Since the observed value of the test statistic falls in the rejection region ($\chi^2 = 1037.833 > 11.0705$), H_0 is rejected. There is sufficient evidence to indicate the proportions of printed invoices in the six error categories differ from the proportions using the previous format at $\alpha = .05$.

b. The observed significance level is p-value $= P(\chi^2 \geq 1037.833)$.

Using Table VII, Appendix A, with df $= 5$, $P(\chi^2 \geq 1037.833)$ $=$ p-value $< .005$

9.43 a. H_0: The row and column classifications are independent
H_a: The row and column classifications are dependent

b. The test statistic is $\chi^2 = \sum \frac{[n_{ij} - \hat{E}(n_{ij})]^2}{\hat{E}(n_{ij})}$

The rejection region requires $\alpha = .01$ in the upper tail of the χ^2 distribution with df = (r - 1)(c - 1) = (2 - 1)(3 - 1) = 2. From Table VII, Appendix A, $\chi^2_{.01} = 9.21034$. The rejection region is $\chi^2 > 9.21034$.

c. Some preliminary calculations are:

$$\hat{E}(n_{11}) = \frac{r_1c_1}{n} = \frac{95(25)}{165} = 14.394 \qquad \hat{E}(n_{21}) = \frac{r_2c_1}{n} = \frac{70(25)}{165} = 10.606$$

$$\hat{E}(n_{12}) = \frac{r_1c_2}{n} = \frac{95(62)}{165} = 35.697 \qquad \hat{E}(n_{22}) = \frac{r_2c_2}{n} = \frac{70(62)}{165} = 26.303$$

$$\hat{E}(n_{13}) = \frac{r_1c_3}{n} = \frac{95(78)}{165} = 44.909 \qquad \hat{E}(n_{23}) = \frac{r_2c_3}{n} = \frac{70(78)}{165} = 33.091$$

d. The test statistic is $\chi^2 = \sum \frac{[n_{ij} - \hat{E}(n_{ij})]^2}{\hat{E}(n_{ij})} = \frac{(10 - 14.394)^2}{14.394}$

$$+ \frac{(32 - 35.697)^2}{35.697} + \frac{(53 - 44.909)^2}{44.909} + \frac{(15 - 10.606)^2}{10.606}$$

$$+ \frac{(30 - 26.303)^2}{26.303} + \frac{(25 - 33.091)^2}{33.091} = 7.50$$

Since the observed value of the test statistic does not fall in the rejection region ($\chi^2 = 7.50 \not> 9.21034$), H_0 is not rejected. There is insufficient evidence to indicate the row and column classifications are dependent at $\alpha = .01$.

9.45 Some preliminary calculations are:

$$\hat{E}(n_{11}) = \frac{r_1c_1}{n} = \frac{156(132)}{437} = 47.121 \qquad \hat{E}(n_{21}) = \frac{184(132)}{437} = 55.579$$

$$\hat{E}(n_{12}) = \frac{156(164)}{437} = 58.545 \qquad \hat{E}(n_{22}) = \frac{184(164)}{437} = 69.053$$

$$\hat{E}(n_{13}) = \frac{156(141)}{437} = 50.334 \qquad \hat{E}(n_{23}) = \frac{184(141)}{437} = 59.368$$

$$\hat{E}(n_{31}) = \frac{97(132)}{437} = 29.300 \qquad \hat{E}(n_{33}) = \frac{97(141)}{437} = 31.297$$

$$\hat{E}(n_{32}) = \frac{97(164)}{437} = 36.403$$

To determine if the row and column classifications are dependent, we test:

H_0: The row and column classifications are independent
H_0: The row and column classifications are dependent

The test statistic is $\chi^2 = \sum \frac{[n_{ij} - \hat{E}(n_{ij})]^2}{\hat{E}(n_{ij})} = \frac{(39 - 47.121)^2}{47.121}$

$$+ \frac{(75 - 58.545)^2}{58.545} + \frac{(42 - 50.334)^2}{50.334} + \frac{(63 - 55.579)^2}{55.579}$$

$$+ \frac{(51 - 69.053)^2}{69.053} + \frac{(70 - 59.368)^2}{59.368} + \frac{(30 - 29.3)^2}{29.3}$$

$$+ \frac{(38 - 36.403)^2}{36.403} + \frac{(29 - 31.297)^2}{31.297} = 15.274$$

The rejection region requires $\alpha = .05$ in the upper tail of the χ^2 distribution with df = $(r - 1)(c - 1) = (3 - 1)(3 - 1) = 4$. From Table VII, Appendix A, $\chi^2_{.05} = 9.48773$. The rejection region is $\chi^2 > 9.48773$.

Since the observed value of the test statistic falls in the rejection region ($\chi^2 = 15.274 > 9.48773$), H_0 is rejected. There is sufficient evidence to indicate the row and column classification are dependent at $\alpha = .05$.

9.47 a. Suppose we select the column classification as the base variable. Next, we must find the percentage of each column that falls in each row category by dividing each cell by the column total and multiplying by 100%. We also divide the row totals by the overall sample size and multiply by 100%.

The cell percentage for A_1B_1 is $\frac{50}{60} \times 100\% = 83.33\%$.

The cell percentage for A_2B_1 is $\frac{10}{60} \times 100\% = 16.67\%$.

The cell percentage for A_1B_2 is $\frac{10}{50} \times 100\% = 20\%$.

The cell percentage for A_2B_2 is $\frac{40}{50} \times 100\% = 80\%$.

The row total percentage for A_1 is $\frac{60}{110} \times 100\% = 54.55\%$

The row total percentage for A_2 is $\frac{50}{110} \times 100\% = 45.45\%$

The table with the percentages is:

		B		
		B_1	B_2	Total
A	A_1	83.33	20	54.55
	A_2	16.67	80	54.45
	Totals	100	100	100

The graph is:

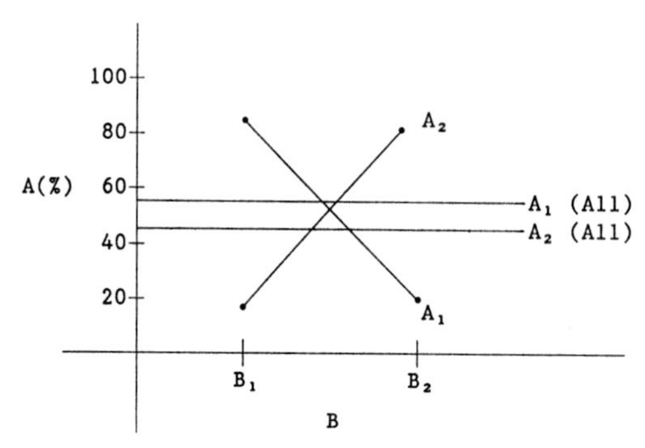

b. We will select the column classification as the base variable. To find the cell percentages, divide the number in each cell by the corresponding column total and multiply by 100%. To find the row percentages, divide the row totals by the total sample size and multiply by 100%.

The percentage for cell C_1D_1 is $\frac{15}{95} \times 100\% = 15.79\%$.

The other percentages are found in a similar manner and are shown in the following table:

		D		
		D_1	D_2	Total
C	C_1	15.79	82.35	47.22
	C_2	84.21	17.65	52.78
	Totals	100	100	100

The graph is:

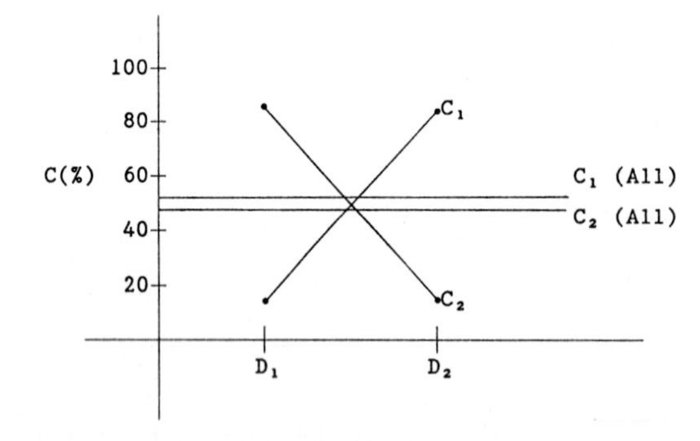

c. We will select the column classification to be the base variable. To find the cell percentages, divide the number in each cell by the corresponding column total and multiply by 100%. To find the row percentages, divide the row totals by the total sample size and multiply by 100%.

The percentage for cell E_1F_1 is $\frac{10}{40} \times 100\% = 25\%$.

The other percentages are found in a similar manner and are shown in the table below:

		F			
		F_1	F_2	F_3	Total
E	E_1	25	75	50	50
	E_2	75	25	50	50
	Total	100	100	100	100

The graph is:

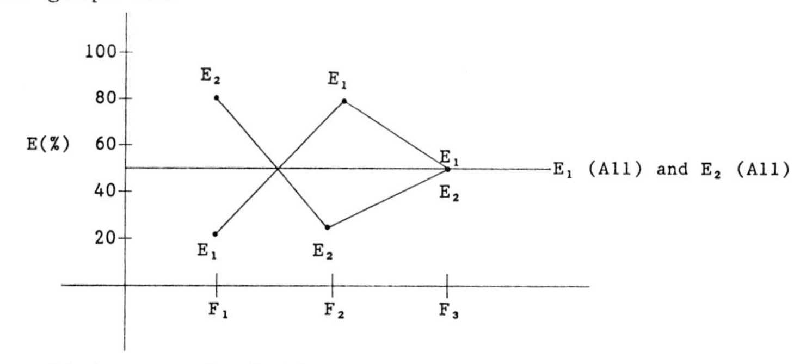

9.49 Some preliminary calculations are:

$$\hat{E}(n_{11}) = \frac{r_1c_1}{n} = \frac{41(28)}{58} = 19.79$$

$$\hat{E}(n_{12}) = \frac{r_1c_2}{n} = \frac{41(30)}{58} = 21.21$$

$$\hat{E}(n_{21}) = \frac{r_2c_1}{n} = \frac{17(28)}{58} = 8.21$$

$$\hat{E}(n_{22}) = \frac{r_2c_2}{n} = \frac{17(30)}{58} = 8.79$$

a. To determine if antisocial behavior and social class are dependent, we test:

H_0: Antisocial behavior and social class are independent
H_0: Antisocial behavior and social class are dependent

The test statistic is $\chi^2 = \sum \frac{[n_{ij} - \hat{E}(n_{ij})]^2}{\hat{E}(n_{ij})} = \frac{(24 - 19.79)^2}{19.79}$

$$+ \frac{(17 - 21.21)^2}{21.21} + \frac{(4 - 8.21)^2}{8.21} + \frac{(13 - 8.79)^2}{8.79} = 5.91$$

The rejection region requires $\alpha = .05$ in the upper tail of the χ^2 distribution with df = (r - 1)(c - 1) = (2 - 1)(2 - 1) = 1. From Table VII, Appendix A, $\chi^2_{.05} = 3.84146$. The rejection region is $\chi^2 > 3.84146$.

Since the observed value of the test statistic falls in the rejection region ($\chi^2 = 5.91 > 3.84146$), H_0 is rejected. There is sufficient evidence to indicate that antisocial behavior and social class are dependent at $\alpha = .05$.

b. p-value = $P(\chi^2 \geq 5.91)$. Using Table VII, Appendix A, with df = 1, $.010 < P(\chi^2 \geq 5.91) < .025$.

The probability of observing our test statistic or anything more unusual if antisocial behavior and social class are independent is between .01 and .025. The author's approximate p-value of .015 is between .01 and .025.

9.51 Some preliminary calculations are:

$$\hat{E}(n_{11}) = \frac{r_1c_1}{n} = \frac{116(80)}{280} = 33.14$$

$$\hat{E}(n_{12}) = \frac{r_1c_2}{n} = \frac{116(100)}{280} = 41.43$$

$$\hat{E}(n_{13}) = \frac{r_1c_3}{n} = \frac{116(100)}{280} = 41.43$$

$$\hat{E}(n_{21}) = \frac{r_2c_1}{n} = \frac{164(80)}{280} = 46.86$$

$$\hat{E}(n_{22}) = \frac{r_2c_2}{n} = \frac{164(100)}{280} = 58.87$$

$$\hat{E}(n_{23}) = \frac{r_2c_3}{n} = \frac{164(100)}{280} = 58.57$$

a. To determine if a relationship exists between marital status and alcoholic classification, we test:

H_0: Marital status and alcoholic classification are independent
H_0: Marital status and alcoholic classification are dependent

The test statistic is $\chi^2 = \sum \frac{[n_{ij} - \hat{E}(n_{ij})]^2}{\hat{E}(n_{ij})} = \frac{(21 - 33.14)^2}{33.14}$

$$+ \frac{(37 - 41.43)^2}{41.43} + \frac{(58 - 41.43)^2}{41.43} + \frac{(59 - 46.86)^2}{46.86}$$

$$+ \frac{(63 - 58.57)^2}{58.57} + \frac{(42 - 58.57)^2}{58.57} = 19.72$$

The rejection region requires $\alpha = .05$ in the upper tail of the χ^2 distribution with df = (r - 1)(c - 1) = (2 - 1)(3 - 1) = 2.

From Table VII, Appendix A, $\chi^2_{.05} = 5.99147$. The rejection region is $\chi^2 > 5.99147$.

Since the observed value of the test statistic falls in the rejection region ($\chi^2 = 19.72 > 5.99147$), H_0 is rejected. There is sufficient evidence to indicate marital status and alcoholic classification are related at $\alpha = .05$.

b. Select alcoholic classification as the base variable. To find the cell percentages, divide the number in each cell by the corresponding column total and multiply by 100%. To find the row percentages, divide the row totals by the total sample size and multiply by 100%.

The percentage for cell "married, diagnosed" is $\frac{21}{80} \times 100\%$ $= 26.25\%$. The remaining percentages are found in a similar manner and are shown in the table below:

		Alcoholic Classification			
		Diagnosed	Undiagnosed	Nonalcoholic	Total
Marital Status	Married	26.25	37	58	41.43
	Not Married	73.75	63	42	58.57
		100	100	100	100

The graph is:

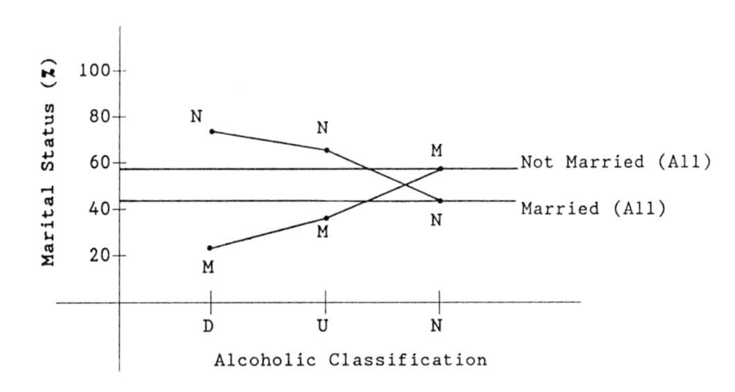

9.53 Some preliminary calculations are:

$$\hat{E}(n_{11}) = \frac{r_1 c_1}{n} = \frac{52(45)}{100} = 23.4$$

$$\hat{E}(n_{12}) = \frac{r_1 c_2}{n} = \frac{52(55)}{100} = 28.6$$

$$\hat{E}(n_{21}) = \frac{r_2 c_1}{n} = \frac{48(45)}{100} = 21.6$$

To determine if a relationship exists between eye and hair color, we test:

H_0: Hair and eye color are independent
H_a: Hair and eye color are dependent

The test statistic is $\chi^2 = \sum \frac{(n_{ij} - \hat{E}(n_{ij}))^2}{\hat{E}(n_{ij})}$

$$= \frac{(31 - 23.4)^2}{23.4} + \frac{(21 - 28.6)^2}{28.6} + \frac{(14 - 21.6)^2}{21.6}$$

$$+ \frac{(34 - 26.4)^2}{26.4} = 9.35$$

The rejection region requires $\alpha = .10$ in the upper tail of the χ^2 distribution with df = (r - 1)(c - 1) = (2 - 1)(2 - 1) = 1. From Table VII, Appendix A, $\chi^2_{.10} = 2.70554$. The rejection region is $\chi^2 > 2.70554$.

Since the observed value of the test statistic falls in the rejection region ($\chi^2 = 9.35 > 2.70554$), H_0 is rejected. There is sufficient evidence to indicate that hair and eye color are related at $\alpha = .10$.

9.55 Some preliminary calculations are:

$\hat{E}(n_{11}) = \frac{r_1c_1}{n} = \frac{36(23)}{41} = 20.195$ $\qquad$ $\hat{E}(n_{12}) = \frac{r_1c_2}{n} = \frac{36(18)}{41} = 15.805$

$\hat{E}(n_{21}) = \frac{r_2c_1}{n} = \frac{5(23)}{41} = 2.805$ $\qquad$ $\hat{E}(n_{22}) = \frac{r_2c_2}{n} = \frac{5(18)}{41} = 2.195$

The test statistic is $\chi^2 = \sum \frac{[n_{ij} - \hat{E}(n_{ij})]^2}{\hat{E}(n_{ij})} = \frac{(21 - 20.195)^2}{20.195}$

$$+ \frac{(15 - 8.05)^2}{15.805} + \frac{(2 - 2.805)^2}{2.805} + \frac{(3 - 2.195)^2}{2.195} = .599$$

The p-value = $P(\chi^2 \geq .599)$ where the χ^2 has df = (r - 1)(c - 1) = (2 - 1)(2 - 1) = 1.

From Table VII, Appendix A,

$.10 < P(\chi^2 \geq .599) < .90$

9.57 a. H_0: $(p_1 - p_2) = 0$
H_a: $(p_1 - p_2) < 0$

The test statistic is $z = \frac{(\hat{p}_1 - \hat{p}_2)}{\sqrt{\frac{p_1q_1}{n_1} + \frac{p_2q_2}{n_2}}} \approx \frac{\hat{p}_1 - \hat{p}_2}{\sqrt{\hat{p}\hat{q}(\frac{1}{n_1} + \frac{1}{n_2})}}$

where $\hat{p}_1 = \frac{x_1}{n_1} = \frac{110}{200} = .55$, $\hat{p}_2 = \frac{x_2}{n_2} = \frac{130}{200} = .65$, and

$$\hat{p} = \frac{x_1 + x_2}{n_1 + n_2} = \frac{110 + 130}{200 + 200} = \frac{240}{400} = .60$$

$$z \approx \frac{.55 - .65}{\sqrt{.6(1 - .6)\left(\frac{1}{200} + \frac{1}{200}\right)}} = \frac{-.10}{.049} = -2.04$$

The rejection region requires $\alpha = .10$ in the lower tail of the z distribution. From Table III, Appendix A, $z_{.10} = 1.28$. The rejection region is $z < -1.28$.

Since the observed value of the test statistic falls in the rejection region ($z = -2.04 < -1.28$), H_0 is rejected. There is sufficient evidence to conclude $(p_1 - p_2) < 0$ at $\alpha = .10$.

b. For confidence coefficient .95, $\alpha = 1 - .95 = .05$ and $\alpha/2 = .05/2 = .025$. From Table III, Appendix A, $z_{.025} = 1.96$. The 95% confidence interval for $(p_1 - p_2)$ is approximately:

$$(\hat{p}_1 - \hat{p}_2) \pm z_{\alpha/2}\sqrt{\frac{\hat{p}_1\hat{q}_1}{n_1} + \frac{\hat{p}_2\hat{q}_2}{n_2}}$$

$$\Rightarrow (.55 - .65) \pm 1.96\sqrt{\frac{.55(1 - .55)}{200} + \frac{.65(1 - .65)}{200}}$$

$$\Rightarrow -.10 \pm .096$$

$$\Rightarrow (-.196, -.004)$$

c. From part (b), $z_{.025} = 1.96$. Using the information from our samples, we can use $p_1 = .55$ and $p_2 = .65$.

$$n_1 = n_2 = \frac{4(z_{\alpha/2})^2(p_1q_1 + p_2q_2)}{W^2}$$

$$= \frac{4(1.96)^2(.55(1 - .55) + .65(1 - .65))}{.01^2}$$

$$= \frac{7.29904}{.0001} = 72990.4 \approx 72,991$$

9.59 a. Some preliminary calculations are:

If all the categories are equally likely,

$$p_{1,0} = p_{2,0} = p_{3,0} = p_{4,0} = p_{5,0} = .2$$

$$E(n_1) = E(n_2) = E(n_3) = E(n_4) = E(n_5) = np_{1,0} = 150(.2) = 30$$

To determine if the categories are not equally likely, we test:

H_0: $p_1 = p_2 = p_3 = p_4 = p_5 = .2$
H_a: At least one probability is different from .2.

The test statistic is $\chi^2 = \sum \frac{[n_i - E(n_i)]^2}{E(n_i)}$

$$= \frac{(28 - 30)^2}{30} + \frac{(35 - 30)^2}{30} + \frac{(33 - 30)^2}{30} + \frac{(25 - 30)^2}{30}$$

$$+ \frac{(29 - 30)^2}{30} = 2.133$$

The rejection region requires $\alpha = .10$ in the upper tail of the χ^2 distribution with df = k - 1 = 5 - 1 = 4. From Table VII, Appendix A, $\chi^2_{.10} = 7.77944$. The rejection region is $\chi^2 > 7.77944$.

Since the observed value of the test statistic does not fall in the rejection region ($\chi^2 = 2.133 \not> 7.77944$), H_0 is not rejected. There is insufficient evidence to indicate the categories are not equally likely at $\alpha = .10$.

b. $\hat{p}_2 = \frac{35}{150} = .233$

For confidence coefficient .90, $\alpha = .10$ and $\alpha/2 = .05$. From Table III, Appendix A, $z_{.05} = 1.645$. The confidence interval is:

$$\hat{p}_2 \pm z_{.05}\sqrt{\frac{\hat{p}_2\hat{q}_2}{n_2}} \Rightarrow .233 \pm 1.645\sqrt{\frac{.233(.767)}{150}}$$

$$\Rightarrow .233 \pm .057 \Rightarrow (.176, .290)$$

9.61 a. Let p_1 = proportion of those that regularly listen to the station before the promotion and p_2 = proportion of those that regularly listen to the station after the promotion. For confidence coefficient .90, $\alpha = 1 - .90 = .10$ and $\alpha/2$ $.10/2 = .05$. From Table III, Appendix A, $z_{.05} = 1.645$. The 90% confidence interval for $p_1 - p_2$ is:

$$(\hat{p}_1 - \hat{p}_2) \pm z_{\alpha/2}\sqrt{\frac{p_1q_1}{n_1} + \frac{p_2q_2}{n_2}}$$

$$\approx (\hat{p}_1 - \hat{p}_2) \pm z_{\alpha/2}\sqrt{\frac{\hat{p}_1\hat{q}_1}{n_1} + \frac{\hat{p}_2\hat{q}_2}{n_2}}$$

where $\hat{p}_1 = \frac{65}{300} = .22$ and $\hat{p}_2 = \frac{154}{500} = .31$

$$\Rightarrow (.22 - .31) \pm 1.645\sqrt{\frac{.22(1 - .22)}{300} + \frac{.31(1 - .31)}{500}}$$

$$\Rightarrow -.09 \pm .05$$

$$\Rightarrow (-.14, -.04)$$

b. For confidence coefficient .95, $\alpha = 1 - .95 = .05$ and $\alpha/2 = .05/2 = .025$. From Table III, Appendix A, $z_{.025} = 1.96$. The 95% confidence interval for p_2 is:

$$\hat{p}_2 \pm z_{\alpha/2}\sqrt{\frac{p_2 q_2}{n_2}} \approx \hat{p}_2 \pm z_{\alpha/2}\sqrt{\frac{\hat{p}_2 \hat{q}_2}{n_2}}$$

$$\Rightarrow .31 \pm 1.96\sqrt{\frac{.31(1 - .31)}{500}}$$

$$\Rightarrow .31 \pm .04$$

$$\Rightarrow (.27, .35)$$

9.63 To determine if there is a difference in the proportion of nonmarketable oranges between the two varieties we test:

H_0: $p_1 - p_2 = 0$
H_a: $p_1 - p_2 \neq 0$

where p_1 and p_2 are the proportions of nonmarketable oranges for Valencia and navel oranges, respectively.

Will need to calculate the following:

$$\hat{p}_1 = \frac{30}{850} = .0353 \qquad \hat{p}_2 = \frac{90}{1500} = .0600 \qquad \hat{p} = \frac{30 + 90}{850 + 1500} = .0511$$

Test statistic: $$z = \frac{(\hat{p}_1 - \hat{p}_2)}{\sqrt{\hat{p}\hat{q}(\frac{1}{n_1} + \frac{1}{n_2})}} = \frac{(.0353 - .0600)}{\sqrt{(.0511)(.9489)(\frac{1}{850} + \frac{1}{1500})}}$$

$$= -2.61$$

The rejection region requires $\alpha/2 = .05/2 = .025$ in each tail of the z distribution. From Table III, Appendix A, $z_{.025} = 1.96$. The rejection region is $z < -1.96$ or $z > 1.96$.

Since the observed value of the test statistic falls in the rejection region ($-2.61 < -1.96$), H_0 is rejected. There is sufficient evidence to indicate a difference between the proportions of nonmarketable Valencia and navel oranges at $\alpha = .05$.

9.65 From Exercise 9.64, $\hat{p}_1 = .2$, $\hat{p}_2 = .1$, $n_1 = 387$, and $n_2 = 311$.

$$\hat{p} = \frac{n_1\hat{p}_1 + n_2\hat{p}_2}{n_1 + n_2} = \frac{387(.2) + 311(.1)}{387 + 311} = .1554$$

To determine if a difference exists between the proportions of the two types of executives who do not know how much poor quality costs their company, we test:

H_0: $p_1 - p_2 = 0$
H_a: $p_1 - p_2 \neq 0$

The test statistic is $z = \dfrac{(\hat{p}_1 - \hat{p}_2) - 0}{\sqrt{\hat{p}\hat{q}\left(\frac{1}{n_1} + \frac{1}{n_2}\right)}}$

$$= \frac{(.2 - .1) - 0}{\sqrt{.1554(.8446)\left(\frac{1}{387} + \frac{1}{311}\right)}} = 3.62$$

The rejection region requires $\alpha/2 = .10/2 = .05$ in each tail of the z distribution. From Table III, Appendix A, $z_{.05} = 1.645$. The rejection region is $z < -1.645$ or $z > 1.645$.

Since the observed value of the test statistic falls in the rejection region ($z = 3.62 > 1.645$), H_0 is rejected. There is sufficient evidence to indicate there is a difference in the proportions of the two types of executives who do not know how much poor quality costs their company at $\alpha = .10$.

9.67 Some preliminary calculations are:

If there is no difference among the five location proportions,

$$p_{1,0} = p_{2,0} = p_{3,0} = p_{4,0} = p_{5,0} = .2$$

$$E(n_1) = np_{1,0} = 425(.2) = 85 = E(n_2) = E(n_3) = E(n_4) = E(n_5)$$

To determine if there is a difference among the proportions of transactions assigned to the five memory locations, we test:

H_0: $p_1 = p_2 = p_3 = p_4 = p_5 = .2$
H_a: At least one p_i is different from .2

The test statistic is $\chi^2 = \sum \dfrac{[n_i - E(n_i)]^2}{E(n_i)}$

$$= \frac{(90 - 85)^2}{85} + \frac{(78 - 85)^2}{85} + \frac{(100 - 85)^2}{85} + \frac{(72 - 85)^2}{85} + \frac{(85 - 85)^2}{85} = 5.506$$

The rejection region requires $\alpha = .025$ in the upper tail of the χ^2 distribution with df = k - 1 = 5 - 1 = 4. From Table VII, Appendix A, $\chi^2_{.025} = 11.1433$. The rejection region is $\chi^2 > 11.1433$.

Since the observed value of the test statistic does not fall in the rejection region ($\chi^2 = 5.506 \not> 11.1433$), H_0 is not rejected. There is insufficient evidence to indicate a difference among the proportions of transactions assigned to the five memory locations at $\alpha = .025$.

9.69 a. The parameter of interest is $p_1 - p_2$, the difference in the population percentages of shots made by players A and B respectively.

b. $(\hat{p}_1 - \hat{p}_2) \pm z_{.025}\sqrt{\frac{\hat{p}_1\hat{q}_1}{n_1} + \frac{\hat{p}_2\hat{q}_2}{n_2}}$

$\Rightarrow (.93 - .86) \pm (1.96)\sqrt{\frac{(.93)(.07)}{100} + \frac{(.86)(.14)}{100}}$

$\Rightarrow .07 \pm .08 \Rightarrow (-.01, .15)$

9.71 For 95% confidence, would use $z_{\alpha/2} = z_{.025} = 1.96$ from Table III, Appendix A.

$$n_1 = n_2 = \frac{(z_{\alpha/2})^2(p_1q_1 + p_2q_2)}{B^2} = \frac{(1.96)^2\{(.425)(.575) + (.463)(.537)\}}{.04^2}$$

$$= 1183.7 \approx 1184$$

9.73 a. The contingency table is:

		Committee: Acceptable	Committee: Rejected	Totals
Inspector	Acceptable	101	23	124
	Rejected	10	19	29
	Totals	111	42	153

b. Yes. To plot the percentages, first convert frequencies to percentages by dividing the numbers in each column by the column total and multiplying by 100. Also, divide the row totals by the overall total and multiply by 100.

		Acceptable	Rejected	
Inspector	Acceptable	$\frac{101}{111} \cdot 100 = 90.99\%$	$\frac{23}{42} \cdot 100 = 54.76\%$	$\frac{124}{153} \cdot 100 = 81.05\%$
	Rejected	$\frac{10}{100} \cdot 100 = 9.01\%$	$\frac{19}{42} \cdot 100 = 45.23\%$	$\frac{29}{153} \cdot 100 = 18.95\%$

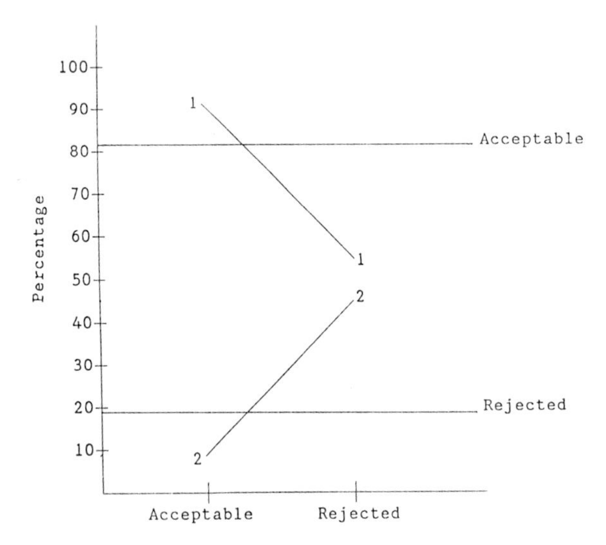

From the plot, it appears there is a relationship.

c. Some preliminary calculations are:

$$\hat{E}(n_{11}) = \frac{r_1 c_1}{n} = \frac{124(111)}{153} = 89.961 \quad \hat{E}(n_{12}) = \frac{r_1 c_2}{n} = \frac{124(42)}{153} = 34.039$$

$$\hat{E}(n_{21}) = \frac{r_2 c_1}{n} = \frac{29(111)}{153} = 21.039 \quad \hat{E}(n_{22}) = \frac{r_2 c_2}{n} = \frac{29(42)}{153} = 7.961$$

To determine if the inspector's classifications and the committee's classifications are related, we test:

H_0: The inspector's and committee's classification are independent

H_a: The inspector's and committee's classifications are dependent

The test statistic is $\chi^2 = \sum\sum \frac{[n_{ij} - \hat{E}(n_{ij})]^2}{\hat{E}(n_{ij})}$

$$= \frac{(101 - 89.961)^2}{89.961} + \frac{(23 - 34.039)^2}{34.039} + \frac{(10 - 21.039)^2}{21.039}$$

$$+ \frac{(19 - 7.961)^2}{7.961} = 26.034$$

The rejection region requires $\alpha = .05$ in the upper tail of the χ^2 distribution with df = $(r - 1)(c - 1) = (2 - 1)(2 - 1) = 1$. From Table VII, Appendix A, $\chi^2_{.05} = 3.84146$. The rejection region is $\chi^2 > 3.84146$.

Since the observed value of the test statistic falls in the rejection region ($\chi^2 = 26.034 > 3.84146$), H_0 is rejected. There is sufficient evidence to indicate the inspector's and committee's classifications are related at $\alpha = .05$. This indicates that the inspector and committee tend to make the same decisions.

CHAPTER

10

SIMPLE LINEAR REGRESSION

10.1 a.

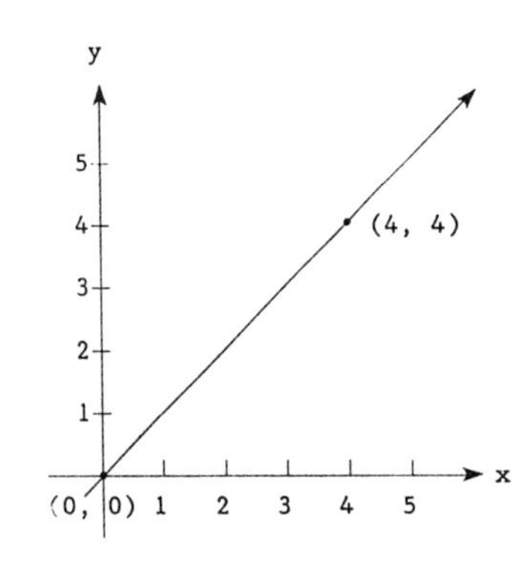

b.

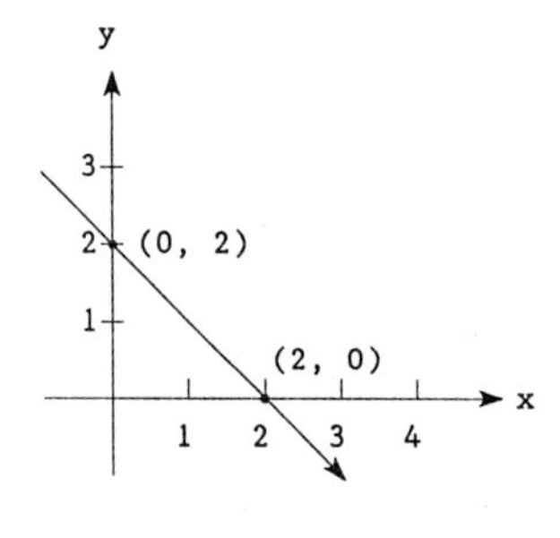

c.

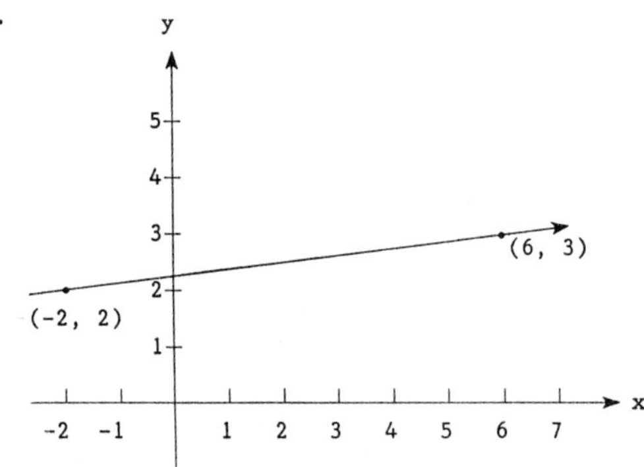

d.

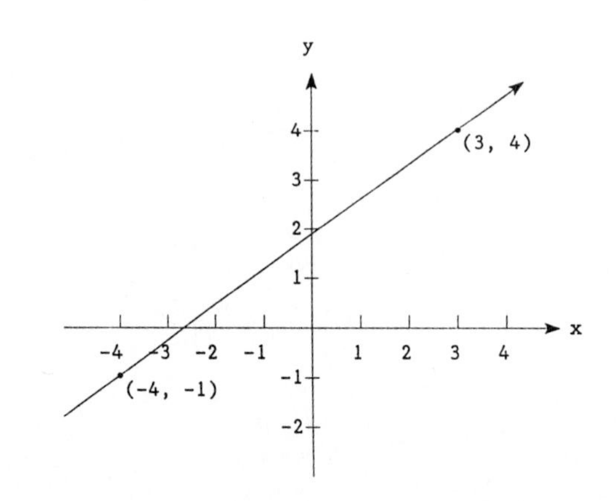

10.3 The 2 equations are

$$3 = \beta_0 + \beta_1(-1) \text{ and } 4 = \beta_0 + \beta_1(3)$$

Subtracting the first equation from the second, we get

$$\begin{array}{rl} 4 = & \beta_0 + 3\beta_1 \\ -(3 = & \beta_0 - \ \beta_1) \\ \hline 1 = & 4\beta_1 \end{array} \quad \Rightarrow \beta_1 = \frac{1}{4}$$

Substituting $\beta_1 = \frac{1}{4}$ into the first equation, we get

$$3 = \beta_0 - \frac{1}{4} \Rightarrow \beta_0 = 3 + \frac{1}{4} = \frac{13}{4}$$

The equation for the line is $y = \frac{13}{4} + \frac{1}{4}x$

10.5 To graph a line, we need 2 points. Pick two values for x, and find the corresponding y values by substituting the values of x into the equation.

a. Let $x = 0 \Rightarrow y = 3 + 2(0) = 3$
and $x = 2 \Rightarrow y + 3 + 2(2) = 7$

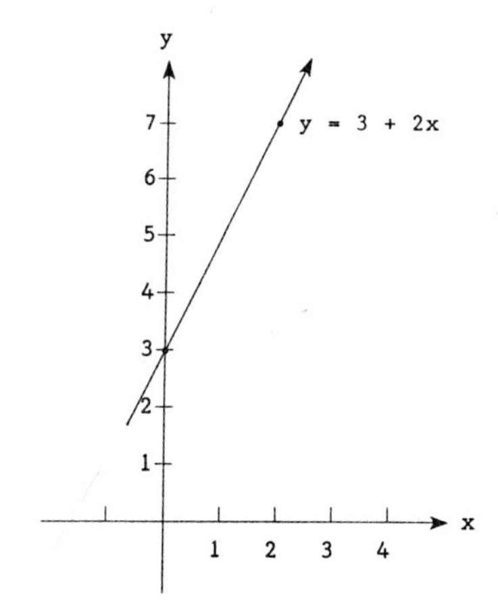

b. Let $x = 0 \Rightarrow y = 3 - 2(0) = 3$
and $x = 2 \Rightarrow y = 3 - 2(2) = -1$

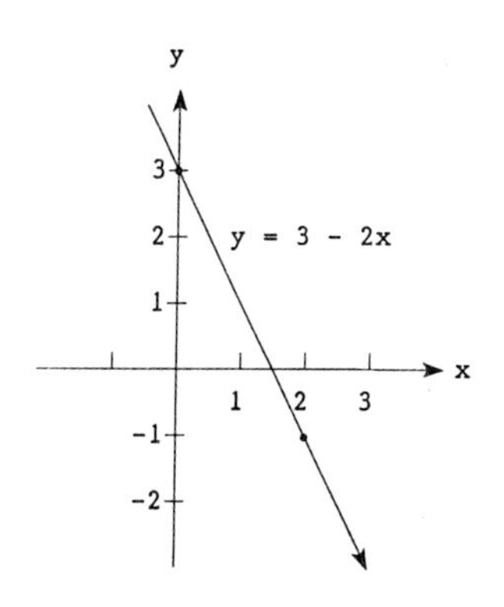

c. Let $x = 0 \Rightarrow y = -3 + 2(0) = -3$
and $x = 2 \Rightarrow y = -3 + 2(2) = 1$

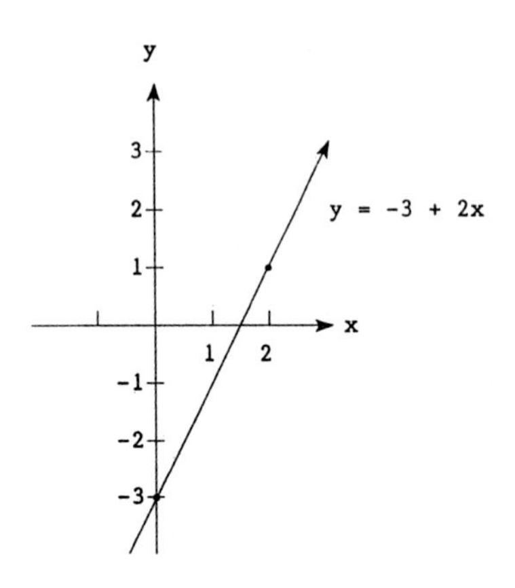

d. Let $x = 0 \Rightarrow y = -0 = 0$
and $x = 2 \Rightarrow y = -2$

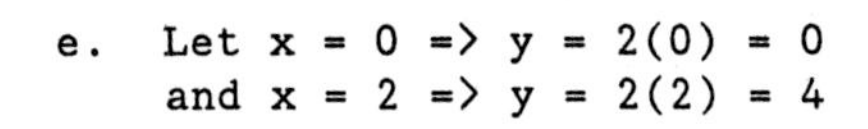

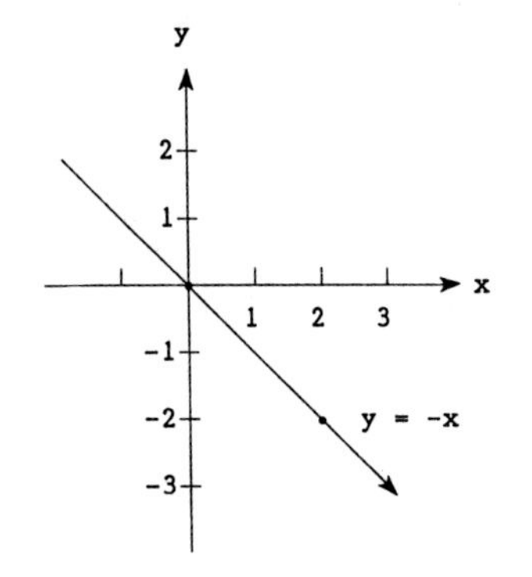

e. Let $x = 0 \Rightarrow y = 2(0) = 0$
and $x = 2 \Rightarrow y = 2(2) = 4$

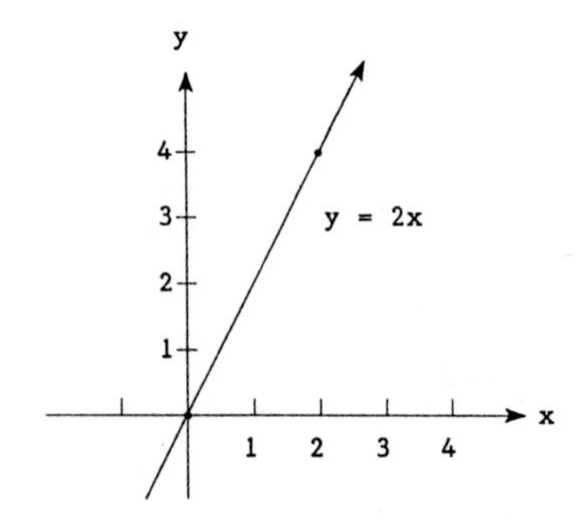

f. Let $x = 0 \Rightarrow y = .5 + 1.25(0) = .5$
and $x = 2 \Rightarrow y = .5 + 1.25(2) = 3$

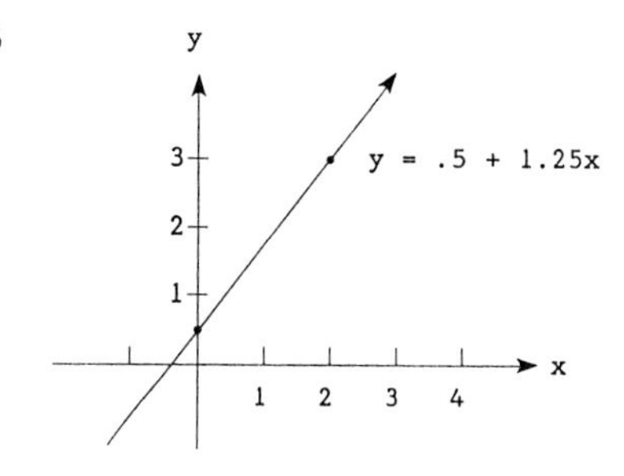

10.7 A deterministic model does not allow for random error or variation, whereas a probabilistic model does. An example where a deterministic model would be appropriate is:

Let y = cost of a 2 x 4 piece of lumber and
x = length (in feet)

The model would be $y = \beta_1 x$. There should be no variation in price for the same length of wood.

An example where a probabilistic model would be appropriate is:

Let y = sales per month of a commodity and
x = amount of money spent advertising.

The model would be $y = \beta_0 + \beta_1 x + \varepsilon$. The sales per month will probably vary even if the amount of money spent on advertising remains the same.

10.9 No. The random error component, ε, allows the values of the variable to fall above or below the line.

10.11 From Exercise 10.10, $\hat{\beta}_0 = 7.10$ and $\hat{\beta}_1 = -.78$.

The fitted line is $\hat{y} = 7.10 - .78x$. To obtain values for $\hat{y}$, we substitute values of x into the equation and solve for $\hat{y}$.

a.

x	y	$\hat{y}=7.10-.78x$	$(y - \hat{y})$	$(y - \hat{y})^2$
7	2	1.64	.36	.1296
4	4	3.98	.02	.0004
6	2	2.42	-.42	.1764
2	5	5.54	-.54	.2916
1	7	6.32	.68	.4624
1	6	6.32	-.32	.1024
3	5	4.76	.24	.0576
			$\sum(y - \hat{y}) = 0.02$	$SSE = \sum(y - \hat{y})^2 = 1.2204$

b.

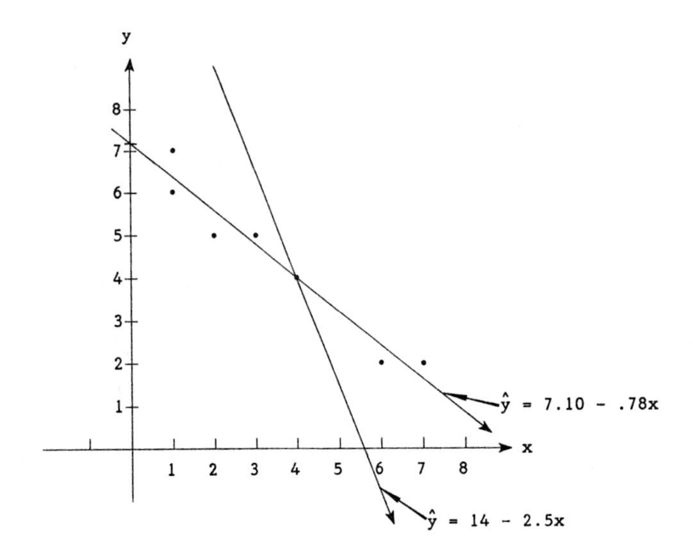

c.

x	y	$\hat{y} = 14-2.5x$	$(y - \hat{y})$	$(y - \hat{y})^2$
7	2	-3.5	5.5	30.25
4	4	4	0	0
6	2	-1	3	9
2	5	9	-4	16
1	7	11.5	-4.5	20.25
1	6	11.5	-5.5	30.25
3	5	6.5	-1.5	2.25
			$\sum(y - \hat{y}) = -7$	SSE = 108.00

10.13 a.

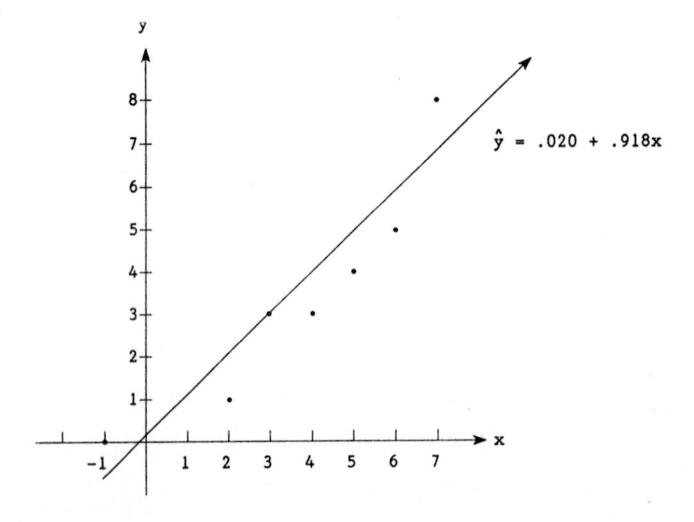

b. As x increases, y tends to increase.

c. $\hat{\beta}_1 = \frac{SS_{xy}}{SS_{xx}} = \frac{39.8571}{43.4286} = .9177616 \approx .918$

$\hat{\beta}_0 = \bar{y} - \hat{\beta}_1\bar{x} = \frac{24}{7} - .9177616\left(\frac{26}{7}\right) = .0197426 \approx .020$

d. The line appears to fit the data quite well.

e. $\hat{\beta}_0 = .020.$ The estimated mean value of y when x = 0 is .020.

$\hat{\beta}_1 = .918.$ The estimated change in the mean value of y for each unit change in x is .918.

These interpretations are valid only for values of x in the range from -1 to 7.

10.15 a.

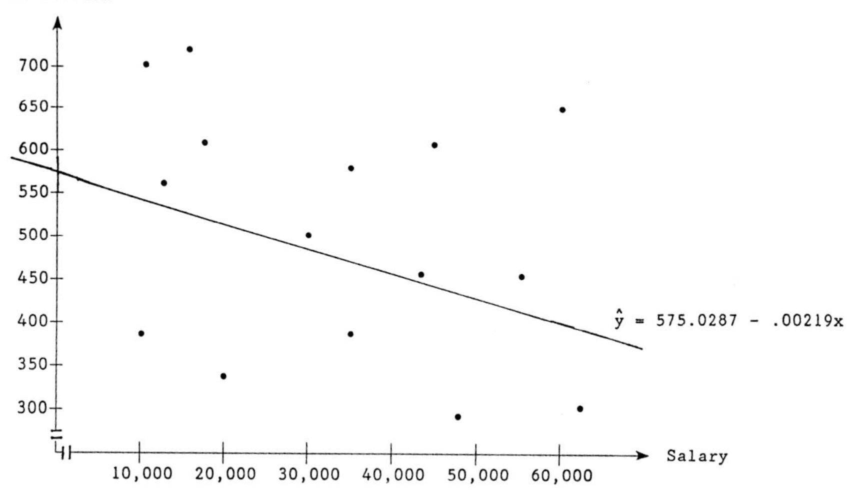

It appears as salary increases, the retaliation index decreases.

b. $\sum x = 516{,}100$ $\sum y = 7497$ $\sum xy = 248{,}409{,}000$

$\sum x^2 = 22{,}119{,}490{,}000$

$\bar{x} = \frac{\sum x}{n} = \frac{516{,}100}{15} = 34{,}406.667$ $\bar{y} = \frac{\sum y}{n} = \frac{7{,}497}{15} = 499.8$

$SS_{xy} = \sum xy - \frac{(\sum x)(\sum y)}{n} = 248{,}409{,}000 - \frac{(516{,}100)(7{,}497)}{15}$

$= 248{,}409{,}000 - 257{,}946{,}780 = -9{,}537{,}780$

$SS_{xx} = \sum x^2 - \frac{(\sum x)^2}{n} = 22{,}119{,}490{,}000 - \frac{(516{,}000)^2}{15}$

$= 22{,}119{,}490{,}000 - 17{,}757{,}281{,}000$

$= 4{,}362{,}209{,}300$

$$\hat{\beta}_1 = \frac{SS_{xy}}{SS_{xx}} = \frac{-9,537,780}{4,362,200,300} = -.00219$$

$$\hat{\beta}_0 = \bar{y} - \hat{\beta}_1\bar{x} = 499.8 - (-.002186456)(34,406.667)$$

$$= 499.8 + 75.22867 = 575.0287$$

$$\hat{y} = 575.0287 - .00219x$$

c. The least squares line supports the answer because the line has a negative slope.

d. $\hat{\beta}_0 = 575.0287$. The estimated mean retailiation index is 575.0287 when the salary is \$0. This is not meaningful because x = 0 is not in the observed range.

e. $\hat{\beta}_1 = -.00219$. When the salary increases by \$1, the mean retailiation index is estimated to decrease by .00219. This is meaningful for the range of x from \$11,900 to \$62,000.

10.17 a. $\sum x_i = 438$ $\quad\sum y_i = 50$ $\quad n = 7$

$\sum x_i^2 = 27,582$ $\quad\sum x_i y_i = 2986$

$$\bar{x} = \frac{\sum x_i}{n} = \frac{438}{7} \qquad \bar{y} = \frac{\sum y_i}{n} = \frac{50}{7}$$

$$SS_{xy} = \sum x_i y_i - \frac{(\sum x_i)(\sum y_i)}{n} = 2986 - \frac{(438)(50)}{7} = -142.57143$$

$$SS_{xx} = \sum x_i^2 - \frac{(\sum x_i)^2}{n} = 27,582 - \frac{(438)^2}{7} = 175.71429$$

$$\hat{\beta}_1 = \frac{SS_{xy}}{SS_{xx}} = \frac{-142.57143}{175.71429} = -.8113821 \approx -.811$$

$$\hat{\beta}_0 = \bar{y} - \hat{\beta}_1\bar{x} = \frac{50}{7} - (-.8113821)(\frac{438}{7}) = 57.912194 \approx 57.912$$

The least squares equation is: $\hat{y} = 57.912 - .811x$

b.

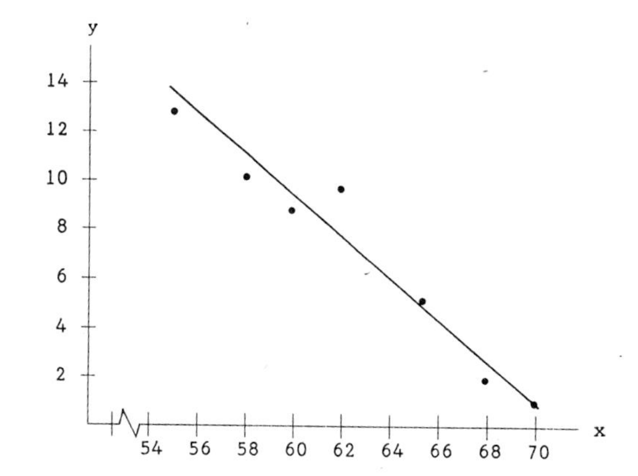

c. If the water temperature is 60°F, the prediction of the drop in pulse rate for a 6-year-old child is:

$$\hat{y} = 57.912 - .811(60) = 9.252 \text{ beats per minute}$$

10.19 a.

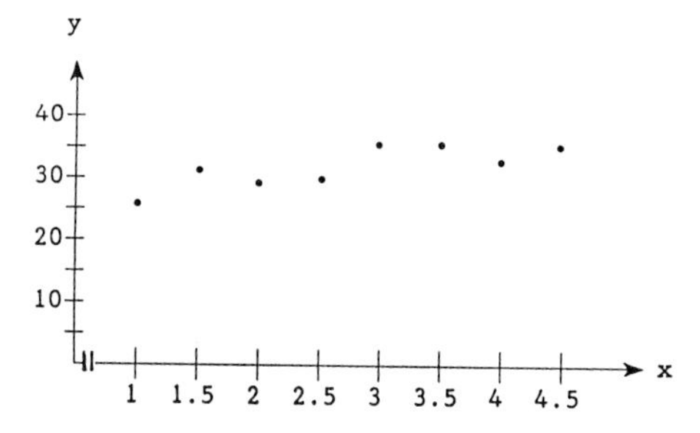

b. $\sum x_i = 22$ $\sum y_i = 248$ $\sum x_i^2 = 71$ $\sum x_i y_i = 707$

$$SS_{xy} = \sum x_i y_i - \frac{(\sum x_i)(\sum y_i)}{n} = 707 - \frac{(22)(248)}{8}$$

$$= 707 - 682 = 25$$

$$SS_{xx} = \sum x_i^2 - \frac{(\sum x_i)^2}{n} = 71 - \frac{(22)^2}{8} = 71 - 60.5 = 10.5$$

$$\hat{\beta}_1 = \frac{SS_{xy}}{SS_{xx}} = \frac{25}{10.5} = 2.3809524 \approx 2.38$$

$$\hat{\beta}_0 = \bar{y} - \hat{\beta}_1\bar{x} = \frac{248}{8} - 2.3809524\left(\frac{22}{8}\right)$$

$$= 31 - 6.5476191 \approx 24.45$$

c. The least squares line is $\hat{y} = 24.45 + 2.38x$. Therefore, if $x = 3.75$, then $\hat{y} = 24.45 + 2.38(3.75) = 24.45 + 9.25 = 33.38$.

10.21 a. $\sum x_i = 35$ $\quad \sum y_i = 1542$ $\quad n = 7$

$\sum x_i^2 = 203$ $\quad \sum x_i y_i = 8068$

$\bar{x} = \frac{\sum x}{n} = \frac{35}{7} = 5 \qquad \bar{y} = \frac{\sum y}{n} = \frac{1542}{7} = 220.286$

$$SS_{xy} = \sum x_i y_i - \frac{(\sum x_i)(\sum y_i)}{n} = 8068 - \frac{(35)(1542)}{7} = 358$$

$$SS_{xx} = \sum x_i^2 - \frac{(\sum x_i)^2}{n} = 203 - \frac{(35)^2}{7} = 28$$

$$\hat{\beta}_1 = \frac{SS_{xy}}{SS_{xx}} = \frac{358}{28} = 12.7857143 \approx 12.786$$

$$\hat{\beta}_0 = \bar{y} - \hat{\beta}_1\bar{x} = 220.286 - 12.7857143(5) = 156.3574285 \approx 156.357$$

The least squares equation is: $\hat{y} = 156.357 + 12.786x$

b.

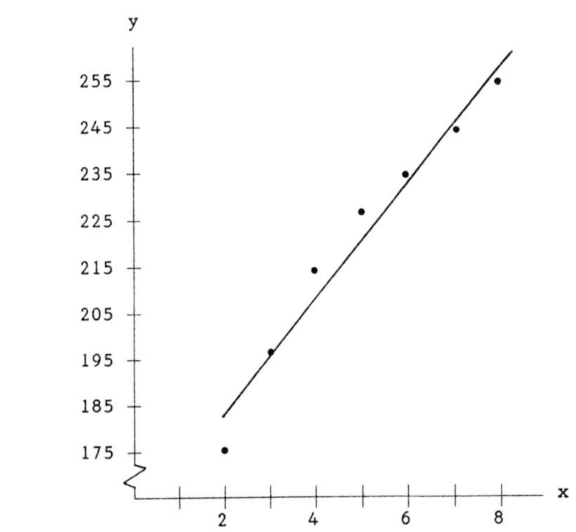

c. $\hat{\beta}_0 = 156.357$. Since x = age = 0 is not in the observed range, $\hat{\beta}_0$ has no meaning.

$\hat{\beta}_1 = 12.786$. For each additional year, the mean length of the sand lance will increase by an estimated 12.786 millimeters.

10.23 a. $s^2 = \frac{SSE}{n - 2} = \frac{.429}{7} = .0613$

b. We would expect most of the observations to be within 2s of the least squares line, This is:

$$2s = 2\sqrt{.0613} \approx .495$$

10.25 $SSE = SS_{yy} - \hat{\beta}_1 SS_{xy}$

where $SS_{yy} = \sum y_i^2 - \dfrac{(\sum y_i)^2}{n}$

For Exercise 10.10,

$$\sum y_i^2 = 159 \qquad \sum y_i = 31 \qquad SS_{yy} = 159 - \frac{31^2}{7} = 159 - 137.285714 = 21.714286$$

$$SS_{xy} = 26.285714 \qquad \hat{\beta}_1 = -.77966$$

Therefore, $SSE = 21.714286 - (-.77966)(-26.285714)$

$$= 21.714286 - 20.49392 = 1.2203662 \approx 1.2204$$

$$s^2 = \frac{SSE}{n-2} = \frac{1.2203662}{5} = .2441, \quad s = \sqrt{.2441} = .4940$$

For Exercise 10.13,

$$\sum x_i = 26 \quad \sum y_i = 24 \quad \sum x_i y_i = 129 \quad \sum x_i^2 = 140 \quad \sum y_i^2 = 124$$

$$SS_{xy} = \sum x_i y_i - \frac{(\sum x_i \sum y_i)}{n} = 129 - \frac{(26)(24)}{7} = 129 - 89.142857 = 39.857143$$

$$SS_{xx} = \sum x_i^2 - \frac{(\sum x_i)^2}{n} = 140 - \frac{(26)^2}{7} = 140 - 96.571429 = 43.428571$$

$$S_{yy} = \sum y_i^2 - \frac{(\sum y_i)^2}{n} = 124 - \frac{(24)^2}{7} = 124 - 82.285714 = 41.714286$$

$$\hat{\beta}_1 = \frac{SS_{xy}}{S_{xx}} = \frac{39.857143}{43.428571} = .9177632$$

$$SSE = SS_{yy} - \hat{\beta}_1 SS_{xy} = 41.714286 - (.9177632)(39.857143)$$

$$= 41.714286 - 36.579418$$

$$= 5.1348681 \approx 5.1349$$

$$s^2 = \frac{SSE}{n-2} = \frac{5.1348681}{5} = 1.0270 \qquad s = \sqrt{1.0270} = 1.0134$$

10.27 a. First, we compute percentage of channels watched by dividing the number of channels watched by the number of channels available and multiplying by 100. The percentages are:

Household	% Channels Watched	Household	% Channels Watched
1	50	11	40
2	34.5	12	75
3	75	13	80
4	40	14	40
5	30	15	56.25
6	60	16	100
7	83.3	17	20
8	100	18	28.9
9	57.1	19	14.3
10	30	20	20

$\sum x_i = 357$ $\quad \sum x_i^2 = 10535$ $\quad \sum x_i y_i = 13100.7$

$\sum y_i = 1034.35$ $\quad \sum y_i^2 = 66743.3125$

$$SS_{xy} = \sum x_i y_i - \frac{\sum x_i \sum y_i}{n} = 13100.7 - \frac{357(1034.35)}{20} = -5362.4475$$

$$SS_{xx} = \sum x_i^2 - \frac{(\sum x_i)^2}{n} = 10535 - \frac{357^2}{20} = 4162.55$$

$$SS_{yy} = \sum y_i^2 - \frac{(\sum y_i)^2}{n} = 66743.3125 - \frac{1034.35^2}{20} = 13249.3164$$

$$\hat{\beta}_1 = \frac{SS_{xy}}{SS_{xx}} = \frac{-5362.4475}{4162.55} = -1.2882602$$

$$\hat{\beta}_0 = \bar{y} - \hat{\beta}_1 \bar{x} = \frac{1034.35}{20} - (-1.2882602)\left(\frac{357}{20}\right) = 74.712945$$

The least squares line is $\hat{y} = 74.7129 - 1.2883x$. Because our estimate for β_1 is negative, -1.2883, this supports the Nielsen findings.

b.

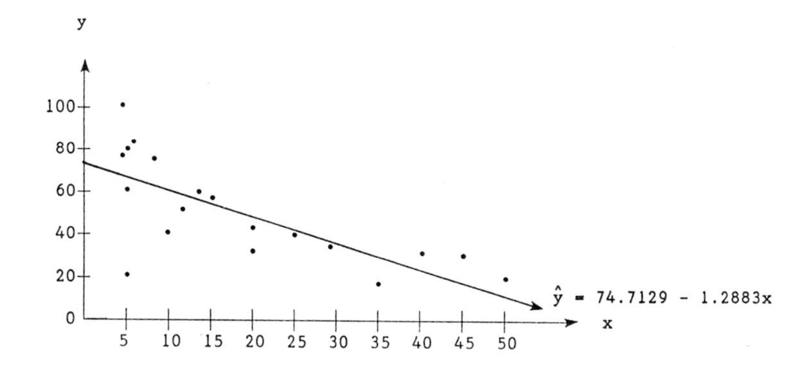

c. $SSE = SS_{yy} - \hat{\beta}_1 SS_{xy} = 13249.3164 - (-1.2882602)(-5362.4475)$
$= 6341.0883$

$$s^2 = \frac{SSE}{n - 2} = \frac{6341.0883}{20 - 2} = 352.2827$$

$$s = \sqrt{352.2827} = 18.769$$

We would expect the least squares line to be able to predict the percentage of channels watched to within 2 standard deviations or $\pm 2(18.769) = \pm 37.538$.

10.29 a.

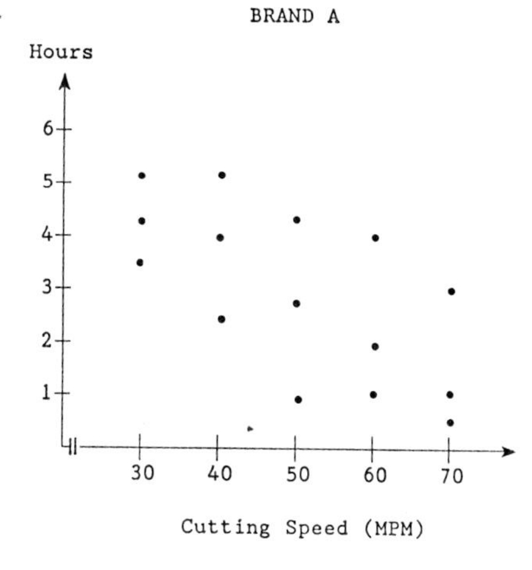

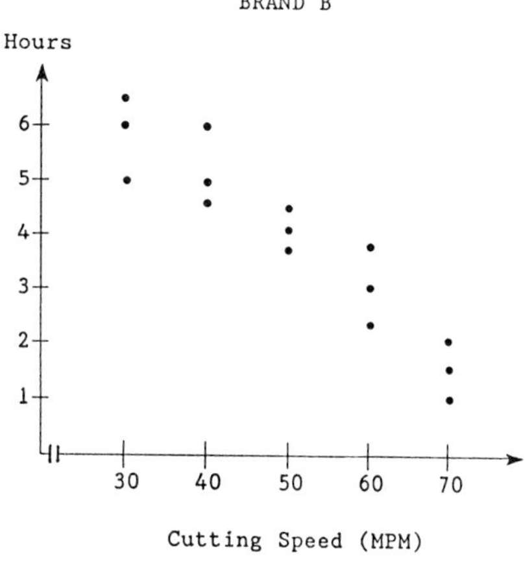

b. For Brand A,

$\sum x_i = 750 \quad \sum y_i = 44.8 \quad \sum x_i y_i = 2022 \quad \sum x_i^2 = 40,500$

$\sum y_i^2 = 168.7$

$$SS_{xx} = \sum x_i^2 - \frac{(\sum x_i)^2}{n} = 40,500 - \frac{750^2}{15} = 40,500 - 37,500 = 3000$$

$$SS_{xy} = \sum x_i y_i - \frac{(\sum x_i)(\sum y_i)}{n} = 2022 - \frac{(750)(44.8)}{15} = 2022 - 2240 = -218$$

$$\hat{\beta}_1 = \frac{SS_{xy}}{SS_{xx}} = \frac{-218}{3000} = -.0726667$$

$$\hat{\beta}_0 = \bar{y} - \hat{\beta}_1 \bar{x} = \frac{44.8}{15} - (-.0726667)\left(\frac{750}{15}\right) = 2.9866667 + 3.633333 = 6.62$$

$$\hat{y} = 6.62 - .0727x$$

For Brand B,

$\sum x_i = 750$ $\sum y_i = 58.9$ $\sum x_i y_i = 2622$ $\sum x_i^2 = 40,500$

$\sum y_i^2 = 270.89$

$$SS_{xx} = \sum x_i^2 - \frac{(\sum x_i)^2}{n} = 40,500 - \frac{(750)^2}{15} = 40,500 - 37,500 = 3000$$

$$SS_{xy} = \sum xy - \frac{(\sum x)(\sum y)}{n} = 2622 - \frac{(750)(58.9)}{15} = 2622 - 2945 = -323$$

$$\hat{\beta}_1 = \frac{SS_{xy}}{SS_{xx}} = \frac{-323}{3000} = -.1076667$$

$$\hat{\beta}_0 = \bar{y} - \hat{\beta}_1 \bar{x} = \left(\frac{59.9}{15}\right) - (-.1076667)\left(\frac{750}{15}\right) = 3.92667 + 5.38333 = 9.31$$

$$\hat{y} = 9.31 - .1077x$$

c. For Brand A,

$$SS_{yy} = \sum y_i^2 - \frac{(\sum y_i)^2}{n} = 168.7 - \frac{(44.8)^2}{15} = 168.7 - 133.80267 = 34.897333$$

$$SSE = SS_{yy} - \hat{\beta}_1 SS_{xy} = 34.897333 - (-.0726667)(-218) = 34.897333 - 15.841341 = 19.055992 \approx 19.056$$

$$s^2 = \frac{SSE}{n-2} = \frac{19.055992}{13} = 1.4658 \qquad s = \sqrt{1.4658} = 1.211$$

For Brand B,

$$SS_{yy} = \sum y_i^2 - \frac{(\sum y_i)^2}{n} = 270.89 - \frac{(58.9)^2}{15} = 270.89 - 231.280667 = 39.609333$$

$$SSE = SS_{yy} - \hat{\beta}_1 SS_{xy} = 39.609333 - (-.1076667)(-323) = 39.609333 - 34.776344 = 4.8329889 \approx 4.833$$

$$s^2 = \frac{SSE}{n-2} = \frac{4.8329889}{13} = .3717 \qquad s = \sqrt{.3717} = .610$$

d. For Brand A,

$\hat{y} = 6.62 - .0727x$. For $x = 70$, $\hat{y} = 6.62 - .0727(70) = 1.531$

$2s = 2(1.211) = 2.422$

Therefore, $\hat{y} \pm 2s \Rightarrow 1.531 \pm 2.422 \Rightarrow (-.891, 3.953)$

For Brand B,

$\hat{y} = 9.31 - .1077x$. For $x = 70$, $\hat{y} = 9.31 - .1077(70) = 1.771$

$2s = 2(.61) = 1.22$

Therefore, $\hat{y} \pm 2s \Rightarrow 1.771 \pm 1.22 \Rightarrow (.551, 2.991)$

e. More confident with Brand B since there is less variation.

10.31 a. For confidence coefficient .95, $\alpha = 1 - .95 = .05$ and $\alpha/2 = .05/2 = .025$. From Table IV, Appendix A, with df $= n - 2 = 10 - 2 = 8$, $t_{.025} = 2.306$.

The 95% confidence interval for β_1 is:

$\hat{\beta}_1 \pm t_{.025} s_{\hat{\beta}_1}$ where $s_{\hat{\beta}_1} = \dfrac{s}{\sqrt{SS_{xx}}} = \dfrac{3}{\sqrt{35}} = .5071$

$\Rightarrow 31 \pm 2.306(.5071) \Rightarrow 31 \pm 1.17 \Rightarrow (29.83, 32.17)$

For confidence coefficient .90, $\alpha = 1 - .90 = .10$ and $\alpha/2 = .10/2 = .05$. From Table IV, Appendix A, with df = 8, $t_{.05} = 1.860$.

The 90% confidence interval for β_1 is:

$\hat{\beta}_1 \pm t_{.05} s_{\hat{\beta}_1}$

$\Rightarrow 31 \pm 1.860(.5071) \Rightarrow 31 \pm .94 \Rightarrow (30.06, 31.94)$

b. $s^2 = \dfrac{SSE}{n - 2} = \dfrac{1,960}{14 - 2} = 163.3333$, $s = \sqrt{s^2} = 12.7802$

For confidence coefficient .95, $\alpha = 1 - .95 = .05$ and $\alpha/2 = .05/2 = .025$. From Table IV, Appendix A, with df $= n - 2 = 14 - 2 = 12$, $t_{.025} = 2.179$. The 95% confidence interval for β_1 is:

$\hat{\beta}_1 \pm t_{.025} s_{\hat{\beta}_1}$ where $s_{\hat{\beta}_1} = \dfrac{s}{\sqrt{SS_{xx}}} = \dfrac{12.7802}{\sqrt{30}} = 2.3333$

$\Rightarrow 64 \pm 2.179(2.3333) \Rightarrow 64 \pm 5.08 \Rightarrow (58.92, 69.08)$

For confidence coefficient .90, $\alpha = 1 - .90 = .10$ and $\alpha/2 = .10/2 = .05$. From Table IV, Appendix A, with df = 12, $t_{.05} = 1.782$.

The 90% confidence interval for β_1 is:

$\hat{\beta}_1 \pm t_{.05} s_{\hat{\beta}_1}$

$\Rightarrow 64 \pm 1.782(2.3333) \Rightarrow 64 \pm 4.16 \Rightarrow (59.84, 68.16)$

c. $s^2 = \frac{SSE}{n - 2} = \frac{146}{20 - 2} = 8.1111$, $s = \sqrt{s^2} = 2.848$

For confidence coefficient .95, $\alpha = 1 - .95 = .05$ and $\alpha/2 = .05/2 = .025$. From Table IV, Appendix A, with df = n - 2 = 20 - 2 = 18, $t_{.025} = 2.776$. The 95% confidence interval for β_1 is:

$$\hat{\beta}_1 \pm t_{.025} s_{\hat{\beta}_1} \text{ where } s_{\hat{\beta}_1} = \frac{s}{\sqrt{SS_{xx}}} = \frac{2.848}{\sqrt{64}} = .356$$

$$\Rightarrow -8.4 \pm 2.101(.356) \Rightarrow -8.4 \pm .75 \Rightarrow (-9.15, -7.65)$$

For confidence coefficient .90, $\alpha = 1 - .90 = .10$ and $\alpha/2 = .10/2 = .05$. From Table IV, Appendix A, with df = 18, $t_{.05} = 1.734$.

The 90% confidence interval for β_1 is:

$$\hat{\beta}_1 \pm t_{.05} s_{\hat{\beta}_1}$$

$$\Rightarrow -8.4 \pm 1.734(.356) \Rightarrow -8.4 \pm .62 \Rightarrow (-9.02, -7.78)$$

10.33 From Exercise 10.32, $\hat{\beta}_1 = .8214$, $s = 1.1922$, $SS_{xx} = 28$, and $n = 7$.

For confidence coefficient .80, $\alpha = 1 - .80 = .20$ and $\alpha/2 = .20/2 = .10$. From Table IV, Appendix A, with df = n - 2 = 7 - 2 = 5, $t_{.10} = 1.476$. The 80% confidence interval for β_1 is:

$$\hat{\beta}_1 \pm t_{.10} s_{\hat{\beta}_1} \text{ where } s_{\hat{\beta}_1} = \frac{s}{\sqrt{SS_{xx}}} = \frac{1.1922}{\sqrt{28}} = .2253$$

$$\Rightarrow .8214 \pm 1.476(.2253) \Rightarrow .8214 \pm .3325 \Rightarrow (.4889, 1.1539)$$

For confidence coefficient .98, $\alpha = 1 - .98 = .02$ and $\alpha/2 = .02/2 = .01$. From Table IV, Appendix A, with df = 5, $t_{.01} = 3.365$.

The 98% confidence interval for β_1 is:

$$\hat{\beta}_1 \pm t_{.01} s_{\hat{\beta}_1}$$

$$\Rightarrow .8214 \pm 3.365(.2253) \Rightarrow .8214 \pm .7581 \Rightarrow (.0633, 1.5795)$$

10.35 a. For n = 100, df = n - 2 = 100 - 2 = 98. From Table IV with df = 98, p-value = $P(t \geq 6.572) + P(t \leq -6.572) = 2P(t \geq 6.572) < 2(.0005) = .0010$. Since this p-value is very small, we would reject H_0. There is sufficient evidence to indicate the slope is not 0. Thus, there is evidence to indicate sales price and square feet of living space are linearly related.

b. For each additional square foot of living space, the price of the house is estimated to increase from \$49.1 to \$90.9. This is valid only for houses with square footage between 1,500 and 4,000. The interval could be made narrower by decreasing the level of confidence.

10.37 a. The probabilistic model is $y = \beta_0 + \beta_1 x + \varepsilon$.

Some preliminary calculations are:

$\sum x = 219$ $\sum x^2 = 6717$ $\sum xy = 13,621$

$\sum y = 505$ $\sum y^2 = 33,405$

$$SS_{xy} = \sum xy - \frac{\sum x \sum y}{n} = 13,621 - \frac{219(505)}{8} = -203.375$$

$$SS_{xx} = \sum x^2 - \frac{(\sum x)^2}{n} = 6717 - \frac{219^2}{8} = 721.875$$

$$SS_{yy} = \sum y^2 - \frac{(\sum y)^2}{n} = 33,405 - \frac{505^2}{8} = 1526.875$$

$$\hat{\beta}_1 = \frac{SS_{xy}}{SS_{xx}} = \frac{-203.375}{721.875} = -.281731601 \approx -.2817$$

$$\hat{\beta}_0 = \bar{y} - \hat{\beta}_1 \bar{x} = \frac{505}{8} - (-.281731601)\left(\frac{219}{8}\right) = 70.8374026 \approx 70.8374$$

The least squares line is $\hat{y} = 70.8 - .28x$

b. Some preliminary calculations are:

$$SSE = SS_{yy} - \hat{\beta}_1 SS_{xy} = 1526.875 - (-.281731601)(-203.375)$$
$$= 1469.577836$$

$$s^2 = \frac{SSE}{n - 2} = \frac{1469.577836}{8 - 2} = 244.9296393, \quad s = \sqrt{244.9296393} = 15.6502$$

To determine if x contributes information for the prediction of y, we test:

$H_0: \beta_1 = 0$
$H_a: \beta_1 \neq 0$

The test statistic is $t = \dfrac{\hat{\beta}_1 - 0}{s_{\hat{\beta}_1}} = \dfrac{-.2817 - 0}{\frac{15.6502}{\sqrt{721.875}}} = -.48$

The rejection region requires $\alpha/2 = .05/2 = .025$ in each tail of the t distribution with df = n - 2 = 8 - 2 = 6. From Table IV, Appendix A, $t_{.025} = 2.447$. The rejection region is $t < -2.447$ or $t > 2.447$.

Since the observed value of the test statistic does not fall in the rejection region ($t = -.48 \not< -2.447$), H_0 is not rejected.

There is insufficient evidence to indicate that x contributes information for the prediction of y at $\alpha = .05$.

10.39 Some preliminary calculations are:

$\sum x = 35$ $\sum x^2 = 203$ $\sum xy = 8068$

$\sum y = 1542$ $\sum y^2 = 344{,}380$

$$SS_{xy} = \sum xy - \frac{\sum x \sum y}{n} = 8068 - \frac{35(1542)}{7} = 358$$

$$SS_{xx} = \sum x^2 - \frac{(\sum x)^2}{n} = 203 - \frac{35^2}{7} = 28$$

$$SS_{yy} = \sum y^2 - \frac{(\sum y)^2}{n} = 344{,}380 - \frac{1542^2}{7} = 4699.4286$$

$$\hat{\beta}_1 = \frac{SS_{xy}}{SS_{xx}} = \frac{358}{28} = 12.78571429$$

$$SSE = SS_{yy} - \hat{\beta}_1 SS_{xy} = 4699.4286 - 12.78571429(358) = 122.142884$$

$$s^2 = \frac{SSE}{n - 2} = \frac{122.142884}{7 - 2} = 24.4285768, \quad s = \sqrt{24.4285768} = 4.9425$$

a. To determine if age, x, contributes information for the prediction of the length, y, of a sand lance, we test:

$H_0: \beta_1 = 0$
$H_a: \beta_1 \neq 0$

The test statistic is $t = \dfrac{\hat{\beta}_1 - 0}{s_{\hat{\beta}_1}} = \dfrac{12.7857}{\frac{4.9425}{\sqrt{28}}} = 13.688$

The rejection region requires $\alpha/2 = .05/2 = .025$ in each tail of the t distribution with df $= n - 2 = 7 - 2 = 5$. From Table IV, Appendix A, $t_{.025} = 2.571$. The rejection region is $t < -2.571$ or $t > 2.571$.

Since the observed value of the test statistic falls in the rejection region ($t = 13.688 > 2.571$), H_0 is rejected. There is sufficient evidence to indicate age contributes information for the prediction of the length of a sand lance at $\alpha = .05$.

b. The p-value is equal to $2P(t \geq 13.69) < 2(.0005) = .001$ from Table IV, Appendix A, with df = 5. For any value of $\alpha > .001$, we would reject H_0.

10.41 From 10.16, $\sum y^2 = 124{,}459$, $\sum y = 1257$, $SS_{xy} = 3472.62439$, and $SS_{xx} = 1098.500771$

$$SS_{yy} = \sum y^2 - \frac{(\sum y)^2}{n} = 124{,}459 - \frac{1257^2}{14} = 11{,}598.3571$$

$$SSE = SS_{yy} - \hat{\beta}_1 SS_{xy} = 11{,}598.3571 - 3.16124(3472.62439) = 620.55797$$

$$s^2 = \frac{SSE}{n-2} = \frac{620.55797}{14-2} = 51.7132, \quad s = \sqrt{51.7132} = 7.1912$$

For confidence coefficient .90, $\alpha = .10$ and $\alpha/2 = .05$. From Table VI, Appendix A, $t_{.05} = 1.782$ with df $= n - 2 = 14 - 2 = 12$. The 90% confidence interval is

$$\hat{\beta}_1 \pm t_{.05} s_{\hat{\beta}_1} \Rightarrow 3.1612 \pm 1.782 \frac{7.1912}{\sqrt{1098.500771}} \Rightarrow 3.1612 \pm .3866$$

$$\Rightarrow (2.7746,\ 3.5478)$$

10.43 From Exercise 10.27, $n = 20$, $SS_{xx} = 4162.55$, $s = 18.769$, and $\hat{\beta}_1 = -1.28826$.

To determine if the percentage of channels watched for 10 minutes or more decreases as the number of channels available increases, we test:

H_0: $\beta_1 = 0$
H_a: $\beta_1 < 0$

The test statistic is $t = \dfrac{\hat{\beta}_1 - 0}{s_{\hat{\beta}_1}} = \dfrac{-1.28826 - 0}{\frac{18.769}{\sqrt{4162.55}}} = -4.43$

The rejection region requires $\alpha = .10$ in the lower tail of the t distribution with df $= n - 2 = 20 - 2 = 18$. From Table IV, Appendix A, $t_{.10} = 1.330$. The rejection region is $t < -1.330$.

Since the observed value of the test statistic falls in the rejection region ($t = -4.43 < -1.330$), H_0 is rejected. There is sufficient evidence to indicate the percentage of channels watched decreases as the number of channels available increases at $\alpha = .10$.

The necessary assumptions are:

1. The mean of the probability distribution of ε is 0.
2. The variance of the probability distribution of ε is constant for all values of x.
3. The probability distribution of ε is normal
4. The errors associated with any 2 different observations are independent.

10.45 a. If $r = .7$, there is a positive relationship between x and y. As x increases, y tends to increase.

b. If $r = -.7$, there is a negative relationship between x and y. As x increases, y tends to decrease.

c. If $r = 0$, there is a 0 slope. There is no relationship between x and y.

d. If $r^2 = .64$, then r is either .8 or −.8. The relationship between x and y could be either positive or negative.

10.47 a. Some preliminary calculations are:

$\sum x = 0$ $\quad \sum x^2 = 10$ $\quad \sum xy = 20$

$\sum y = 12$ $\quad \sum y^2 = 70$

$$SS_{xy} = \sum xy - \frac{\sum x \sum y}{n} = 20 - \frac{0(12)}{5} = 20$$

$$SS_{xx} = \sum x^2 - \frac{(\sum x)^2}{n} = 10 - \frac{0^2}{5} = 10$$

$$SS_{yy} = \sum y^2 - \frac{(\sum y)^2}{n} = 70 - \frac{12^2}{5} = 41.2$$

$$r = \frac{SS_{xy}}{\sqrt{SS_{xx} SS_{yy}}} = \frac{20}{\sqrt{10(41.2)}} = .9853$$

$r^2 = .9853^2 = .9709$

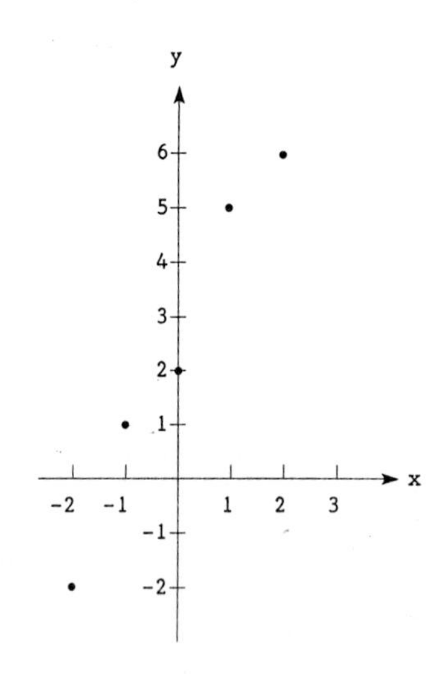

b. Some preliminary calculations are:

$\sum x = 0$ $\quad \sum x^2 = 10$ $\quad \sum xy = -15$

$\sum y = 16$ $\quad \sum y^2 = 74$

$$SS_{xy} = \sum xy - \frac{\sum x \sum y}{n} = -15 - \frac{0(16)}{5} = -15$$

$$SS_{xx} = \sum x^2 - \frac{(\sum x)^2}{n} = 10 - \frac{0^2}{5} = 10$$

$$SS_{yy} = \sum y^2 - \frac{(\sum y)^2}{n} = 74 - \frac{16^2}{5} = 22.8$$

$$r = \frac{SS_{xy}}{\sqrt{SS_{xx}SS_{yy}}} = \frac{-15}{\sqrt{10(22.8)}} = -.9934$$

$$r^2 = (-.9934)^2 = .9868$$

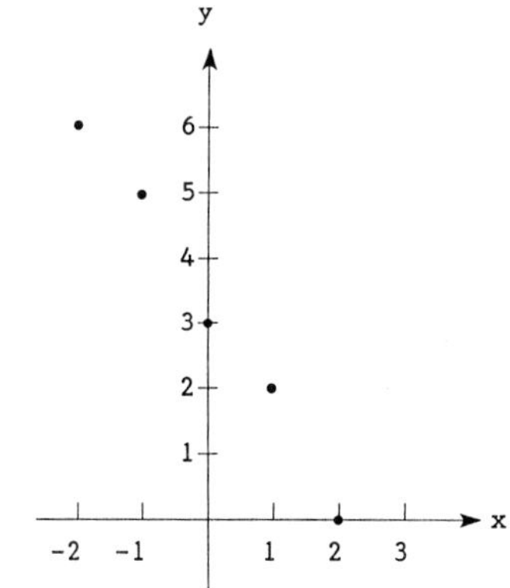

c. Some preliminary calculations are:

$\sum x = 18$ $\quad \sum x^2 = 52$ $\quad \sum xy = 36$

$\sum y = 14$ $\quad \sum y^2 = 32$

$$SS_{xy} = \sum xy - \frac{\sum x \sum y}{n} = 36 - \frac{18(14)}{7} = 0$$

$$SS_{xx} = \sum x^2 - \frac{(\sum x)^2}{n} = 52 - \frac{18^2}{7} = 5.7143$$

$$SS_{yy} = \sum y^2 - \frac{(\sum y)^2}{n} = 32 - \frac{14^2}{7} = 4$$

$$r = \frac{SS_{xy}}{\sqrt{SS_{xx}SS_{yy}}} = \frac{0}{\sqrt{5.7143(4)}} = 0$$

$$r^2 = 0^2 = 0$$

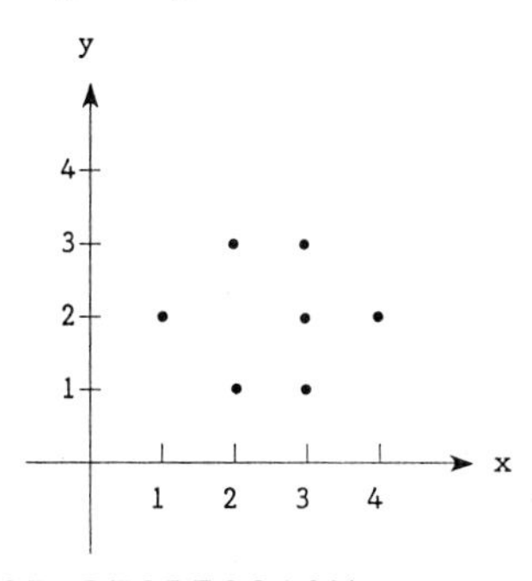

d. Some preliminary calculations are:

$\sum x = 15$ $\sum x^2 = 71$ $\sum xy = 12$

$\sum y = 4$ $\sum y^2 = 6$

$$SS_{xy} = \sum xy - \frac{\sum x \sum y}{n} = 12 - \frac{15(4)}{5} = 0$$

$$SS_{xx} = \sum x^2 - \frac{(\sum x)^2}{n} = 71 - \frac{15^2}{5} = 26$$

$$SS_{yy} = \sum y^2 - \frac{(\sum y)^2}{n} = 6 - \frac{4^2}{5} = 2.8$$

$$r = \frac{SS_{xy}}{\sqrt{SS_{xx}SS_{yy}}} = \frac{0}{\sqrt{26(2.8)}} = 0$$

$$r^2 = 0^2 = 0$$

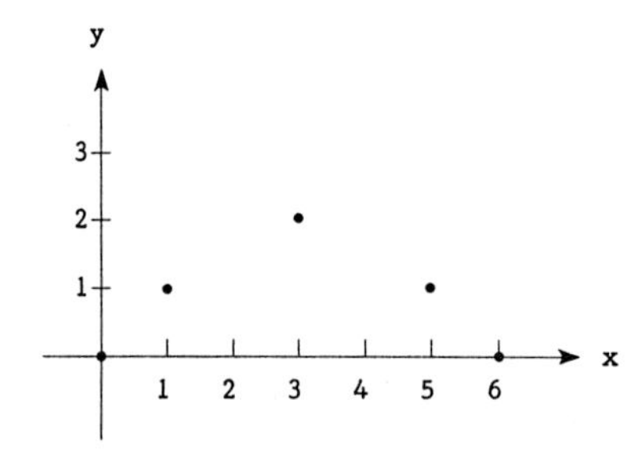

10.49 Some preliminary calculations are:

$\sum x = 72,872$ $\sum x^2 = 466,515,984$ $\sum xy = 648,132.6$

$\sum y = 98.7$ $\sum y^2 = 929.55$

$$SS_{xy} = \sum xy - \frac{\sum x \sum y}{n} = 648,132.6 - \frac{72,872(98.7)}{12} = 48,760.4$$

$$SS_{xx} = \sum x^2 - \frac{(\sum x)^2}{n} = 466,515,984 - \frac{72,872^2}{12} = 23,988,618.7$$

$$SS_{yy} = \sum y^2 - \frac{(\sum y)^2}{n} = 929.55 - \frac{98.7^2}{12} = 117.7425$$

$$r = \frac{SS_{xy}}{\sqrt{SS_{xx}SS_{yy}}} = \frac{48,760.4}{\sqrt{23988618.7(117.7425)}} = .9175$$

$$r^2 = .9175^2 = .8418$$

There is a very strong positive linear relationship, $r = .9175$, between the number of 18-hole or larger golf courses in the U.S. and the number of divorces in the U.S.

r^2 = .8418 => 84.18% of the total sample variability around the mean number of divorces is explained by the linear relationship between the number of golf courses and the number of divorces.

No causal relationship can be inferred. The increase in the number of golf courses has not <u>caused</u> an increase in the number of divorces.

10.51 From Exercise 10.20, $SS_{xy} = 100.0825$ and $SS_{xx} = 83.3425$.

$\sum y = 102.1 \qquad \sum y^2 = 996.21$

$$SS_{yy} = \sum y^2 - \frac{(\sum y)^2}{n} = 996.21 - \frac{102.1^2}{12} = 127.5092$$

$$r = \frac{SS_{xy}}{\sqrt{SS_{xx}SS_{yy}}} = \frac{100.0825}{\sqrt{83.3425(127.5092)}} = .9709$$

$r^2 = .9709^2 = .9426$

r = .9709 implies the cocoon temperature and the outside air temperature are very strongly related. The relationship is positive.

r^2 = .9426 implies that 94.26% of the total sample variability around the mean cocoon temperature is explained by the linear relationship between cocoon temperature and outside air temperature.

10.53 a.

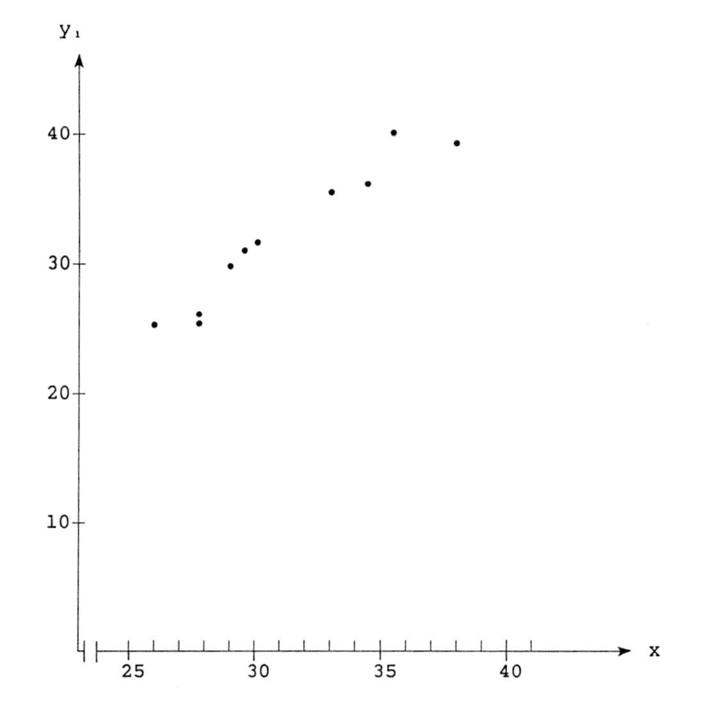

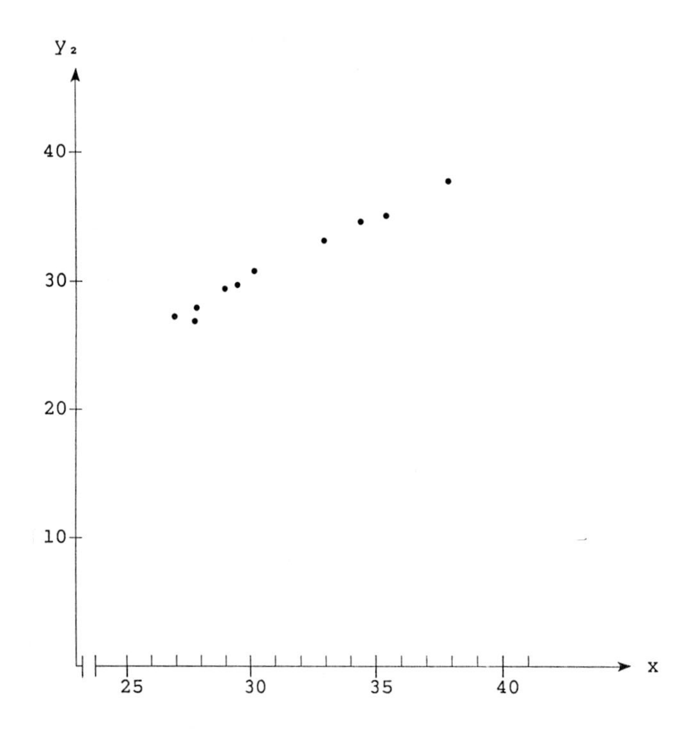

b. It appears that the weigh-in-motion reading after calibration adjustment is more highly correlated with the static weight of trucks than prior to calibration adjustment. The scattergram is closer to a straight line.

c. Some preliminary calculations are:

$\sum x = 312.8$ $\qquad \sum x^2 = 9911.42$ $\qquad \sum xy_1 = 10,201.41$

$\sum y_1 = 320.2$ $\qquad \sum y_1^2 = 10,543.68$

$\sum y_2 = 311.2$ $\qquad \sum y_2^2 = 9809.52$ $\qquad \sum xy_2 = 9859.84$

$$SS_{xy_1} = \sum xy_1 - \frac{\sum x \sum y_1}{n} = 10,201.41 - \frac{312.8(320.2)}{10} = 185.554$$

$$SS_{xx} = \sum x^2 - \frac{(\sum x)^2}{n} = 9911.42 - \frac{312.8^2}{10} = 127.036$$

$$SS_{y_1y_1} = \sum y_1^2 - \frac{(\sum y_1)^2}{n} = 10,543.68 - \frac{320.2^2}{10} = 290.876$$

$$SS_{xy_2} = \sum xy_2 - \frac{\sum x \sum y_2}{n} = 9859.84 - \frac{312.8(311.2)}{10} = 125.504$$

$$SS_{y_2y_2} = \sum y_2^2 - \frac{(\sum y_2)^2}{n} = 9809.52 - \frac{311.2^2}{10} = 124.976$$

$$r_1 = \frac{SS_{xy_1}}{\sqrt{SS_{xx}SS_{y_1y_1}}} = \frac{185.554}{\sqrt{127.036(290.876)}} = .9653$$

$$r_2 = \frac{SS_{xy_2}}{\sqrt{SS_{xx}SS_{y_2y_2}}} = \frac{125.504}{\sqrt{127.036(124.976)}} = .9960$$

$r_1 = .9653$ implies the static weight of trucks and weigh-in-motion prior to calibration adjustment have a strong positive relationship.

$r_2 = .996$ implies the static weight of trucks and weigh-in-motion after calibration adjustment have a stronger positive relationship.

The closer r is to 1 indicates the more accurate the weigh-in-motion readings are.

d. Yes. If the weigh-in-motion readings were all exactly the same distance below (or above) the actual readings, r would be 1.

10.55 a. Some preliminary calculations are:

$\sum x_2 = 10{,}166$ $\quad\sum x_2^2 = 37{,}987{,}830$ $\quad\sum x_2 y = 45{,}986{,}410$

$\sum y = 10{,}595$ $\quad\sum y^2 = 57{,}142{,}963$

$$SS_{x_2y} = \sum x_2 y - \frac{\sum x_2 \sum y}{n} = 45{,}986{,}410 - \frac{10{,}166(10{,}595)}{10} = 35{,}215{,}533$$

$$SS_{x_2x_2} = \sum x_2^2 - \frac{(\sum x_2)^2}{n} = 37{,}987{,}830 - \frac{10{,}166^2}{10} = 27{,}653{,}074.4$$

$$SS_{yy} = \sum y^2 - \frac{(\sum y)^2}{n} = 57{,}142{,}963 - \frac{10{,}595^2}{10} = 45{,}917{,}560.5$$

$$\hat{\beta}_2 = \frac{SS_{x_2y}}{SS_{x_2x_2}} = \frac{35{,}215{,}533}{27{,}653{,}074.4} = 1.273476232 \approx 1.273$$

$$\hat{\beta}_0 = \bar{y} - \hat{\beta}_2\bar{x}_2 = \frac{10{,}595}{10} - 1.273476232\left(\frac{10{,}166}{10}\right) = -235.115937 \approx -235.1$$

The least squares line is $\hat{y} = -235.1 + 1.273x_2$

b. $SSE = SS_{yy} - \hat{\beta}_1 SS_{x_2y} = 45{,}917{,}560.5 - 1.273476232(35{,}215{,}533)$
$= 1{,}071{,}416.23$

$$s^2 = \frac{SSE}{n-2} = \frac{1{,}071{,}416.23}{10-2} = 133{,}927.0288, \quad s = \sqrt{133{,}927.0288} = 365.9604$$

To determine if the number of arrests increases as the number of law enforcement employees increases, we test:

H_0: $\beta_2 = 0$
H_a: $\beta_2 > 0$

The test statistic is $t = \dfrac{\hat{\beta}_2 - 0}{\dfrac{s}{\sqrt{SS_{x_2x_2}}}} = \dfrac{1.273 - 0}{\dfrac{365.9604}{\sqrt{27,653,074.4}}} = 18.29$

The rejection requires $\alpha = .05$ in the upper tail of the t distribution with df = n - 2 = 10 - 2 = 8. From Table IV, Appendix A, $t_{.05} = 1.860$. The rejection region is $t > 1.860$.

Since the observed value of the test statistic falls in the rejection region ($t = 18.29 > 1.86$), H_0 is rejected. There is sufficient evidence to indicate that as the number of law enforcement employees increase, the number of arrests also increase, at $\alpha = .05$.

c. $r^2 = 1 - \dfrac{SSE}{SS_{yy}} = 1 - \dfrac{1,071,416.23}{45,917,560.5} = .9767$

$r^2 = .9767$ implies that 97.67% of the total sample variability around the mean number of arrests is explained by the linear relationship between the number of arrests and the number of law enforcement employees.

d. For Exercise 10.54, $SSE = SS_{yy} - \hat{\beta}_1 SS_{x_1y}$

$= 45,917,560.5 - 6.20330367(4,325,998.5)$

$= 19,082,078.13$

This is much larger than SSE = 1,071,416.23 for this problem.

e. For Exercise 10.54, $r^2 = .5847$ while $r^2 = .9767$ for this problem. Thus, the number of law enforcement employees explains more of the variation in y than does population density.

10.57 a.

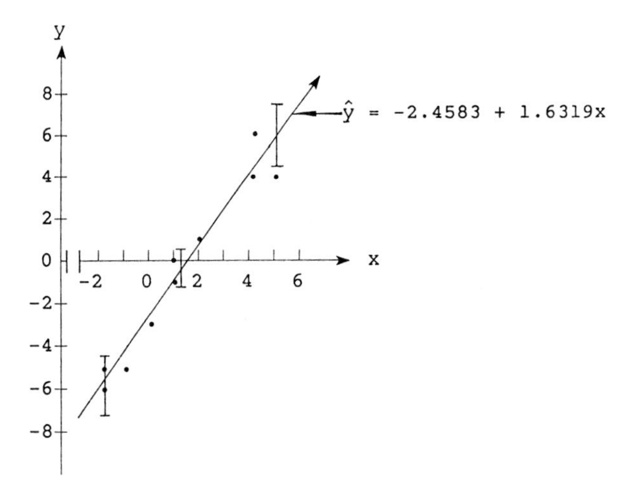

c. Some preliminary calculations are:

$$SSE = SS_{yy} - \hat{\beta}_1 SS_{xy} = 162.5 - 1.631944444(94) = 9.0972$$

$$s^2 = \frac{SSE}{n - 2} = \frac{9.0972}{10 - 2} = 1.137153, \quad s = \sqrt{1.137153} = 1.0664$$

$$\bar{x} = \frac{\sum x}{n} = \frac{12}{10} = 1.2$$

For x = 5, $\hat{y} = -2.4583 + 1.6319(5) = 5.7012$

For confidence coefficient .95, $\alpha = 1 - .95 = .05$ and $\alpha/2 = .05/2 = .025$. From Table IV, Appendix A, $t_{.025} = 2.306$ with df = n - 2 = 10 - 2 = 8. The 95% confidence interval is:

$$\hat{y} \pm t_{\alpha/2} s \sqrt{\frac{1}{n} + \frac{(x_p - \bar{x})^2}{SS_{xx}}}$$

$$\Rightarrow 5.7012 \pm 2.306(1.0664)\sqrt{\frac{1}{10} + \frac{(5 - 1.2)^2}{57.6}}$$

$$\Rightarrow 5.7012 \pm 1.4563 \Rightarrow (4.2449, 7.1575)$$

d. For x = 1.2, $\hat{y} = -2.4583 + 1.6319(1.2) = -.5$. The 95% confidence interval is:

$$-.5 \pm 2.306(1.0664)\sqrt{\frac{1}{10} + \frac{(1.2 - 1.2)^2}{57.6}} \Rightarrow -.5 \pm .7776$$
$$\Rightarrow (-1.2776, .2776)$$

For x = -2, $\hat{y} = -2.4583 + 1.6319(-2) = -5.7221$. The 95% confidence interval is:

$$-5.7221 \pm 2.306(1.0664)\sqrt{\frac{1}{10} + \frac{(-2 - 1.2)^2}{57.6}} \Rightarrow -5.7221 \pm 1.2961$$
$$\Rightarrow (-7.0182, -4.4260)$$

e. The width of the confidence interval when $x = 1.2$ is smaller than that when $x = -2$ or $x = 5$. The widths of the intervals change because of the term $\frac{(x_p - \bar{x})^2}{SS_{xx}}$. If $x_p = \bar{x}$, the term is 0. As x_p moves further away from $\bar{x}$, the width of the interval increases.

10.59 a. From Exercise 10.20, $\hat{y} = 3.375 + 1.201x$. From Exercise 10.38, $s^2 = .73243$, $s = \sqrt{.73243} = .8558$, $\bar{x} = 4.275$, and $SS_{xx} = 83.3425$

For $x = 7$, $\hat{y} = 3.375 + 1.201(7) = 11.782$

For confidence coefficient .95, $\alpha = 1 - .95 = .05$ and $\alpha/2 = .05/2 = .025$. From Table IV, Appendix A, with df $= n - 2 = 12 - 2 = 10$, $t_{.025} = 2.228$. The 95% confidence interval is:

$$\hat{y} \pm t_{\alpha/2} s \sqrt{\frac{1}{n} + \frac{(x_p - \bar{x})^2}{SS_{xx}}}$$

$$\Rightarrow 11.782 \pm 2.228(.8558)\sqrt{\frac{1}{12} + \frac{(7 - 4.275)^2}{83.3425}}$$

$$\Rightarrow 11.782 \pm .7918 \Rightarrow (10.990, 12.574)$$

b. The 95% prediction interval is:

$$\hat{y} \pm t_{\alpha/2} s \sqrt{1 + \frac{1}{n} + \frac{(x_p - \bar{x})^2}{SS_{xx}}}$$

$$\Rightarrow 11.782 \pm 2.228(.8558)\sqrt{1 + \frac{1}{12} + \frac{(7 - 4.275)^2}{83.3425}}$$

$$\Rightarrow 11.782 \pm 2.0646 \Rightarrow (9.7174, 13.8466)$$

10.61 From Exercise 10.21, we have $\hat{y} = 156.357 + 12.786x$
For both parts of this exercise, $\hat{y} = 156.357 + 12.876(4) = 207.5$

We will also need to calculate

$$s = \frac{SSE}{n - 2}$$

Here,

$$SSE = SS_{yy} - \hat{\beta}_1 SS_{xy}$$

$$SS_{yy} = \sum y^2 - \frac{(\sum y)^2}{n} = 344380 - \frac{(1542)^2}{7} = 4699.43$$

$$SSE = 4699.43 - 12.786(358) = 122.04$$

$$s = \frac{122.04}{5} = 4.94$$

a. The form of the confidence interval is

$$\hat{y} \pm t_{\alpha/2} s \sqrt{\frac{1}{n} + \frac{(x_p - \bar{x})^2}{SS_{xx}}}$$

For confidence coefficient .95, α = .05 and $\alpha/2$ = .025. From Table VI, Appendix A, with df = n - 2 = 7 - 2 = 5, $t_{.025}$ = 2.571.

The confidence interval is

$$207.5 \pm 2.571(4.94)\sqrt{\frac{1}{7} + \frac{(4 - 5)^2}{28}}$$

$$\Rightarrow 207.5 \pm 5.37 \Rightarrow (202.13, 212.87)$$

We can be 95% confident that the estimate of the mean length of 4 year old sand lances is between 202.13 and 212.87.

b. The form of the prediction interval is

$$\hat{y} \pm t_{\alpha/2} s \sqrt{1 + \frac{1}{n} + \frac{(x_p - \bar{x})^2}{SS_{xx}}}$$

The same table value of 2.571 from part (a) will be used since the confidence coefficient and degrees of freedom are identical.

The prediction interval is

$$207.5 \pm 2.571(4.94)\sqrt{1 + \frac{1}{7} + \frac{(4 - 5)^2}{28}}$$

$$\Rightarrow 207.5 \pm 13.80 \Rightarrow (193.70, 221.30)$$

We can be 95% confident that the estimate of the length of a 4 year old sand lance is between 193.70 and 221.30. The difference between the two intervals is that the first is for the mean value for 4 year old sand lances, while the second involves individual predictions for 4 year old sand lances.

10.63 From Exercise 10.22, the least squares line is $\hat{y} = -8.368 + 17.612x$ and SS_{xx} = 8.8675. From Exercise 10.42, s^2 = 7.838, $s = \sqrt{7.383}$ = 2.7996, and $\bar{x} = \frac{7.7}{4} = 1.925$.

For x = 2.5, $\hat{y}$ = -8.368 + 17.612(2.5) = 35.662

For confidence coefficient .95, α = 1 - .95 = .05 and $\alpha/2$ = .05/2 = .025. From Table IV, Appendix A, with df = n - 2 = 4 - 2 = 2, $t_{.025}$ = 4.303. The 95% confidence interval is:

$$\hat{y} \pm t_{\alpha/2}\, s\sqrt{\frac{1}{n} + \frac{(x_p - \bar{x})^2}{SS_{xx}}}$$

$$=> 35.662 \pm 4.303(2.7996)\sqrt{\frac{1}{4} + \frac{(2.5 - 1.925)^2}{8.8675}}$$

$$=> 35.662 \pm 6.457 => (29.205,\ 42.119)$$

We are 95% confident that the mean weight gain of rainbow trout is between 29.205% and 42.119% when the ration level is 2.5% of body weight and the water temperature is 12°C.

10.65 a. From the printout, $\hat{\beta}_0 = 575.028672$ and $\hat{\beta}_1 = -.002186$

b. From the printout, SSE = 293,208.461, s^2 = MSE = 22,554.497, and s = Root MSE = 150.18155.

Based on s, we would expect to be able to predict the retaliation index to within approximately ±2(150.18155) or to within ±300.3631 units of its actual value.

c. r^2 = R-Square = .0664 implies 6.64% of the total sample variation around the mean retaliation index is explained by the linear relationship between retaliation index and the salary of the whistle blower.

d. t = -.962 and the observed significance level is .3538. Since the p-value = .3538 is larger than $\alpha = .05$, H_0 is not rejected. There is insufficient evidence to indicate the model is useful.

e. For salary = $35,000, we will use the 6th observation. From the printout, the 95% confidence interval for the mean is (414.7, 582.3).

10.67 a. $\hat{\beta}_1 = \frac{SS_{xy}}{SS_{xx}} = \frac{-30}{50} = -.6$, $\hat{\beta}_0 = \bar{y} - \hat{\beta}_1\bar{x} = 27 - (-.6)(1.3) = 27.78$

The least squares line is $\hat{y} = 27.78 - .6x$

b.

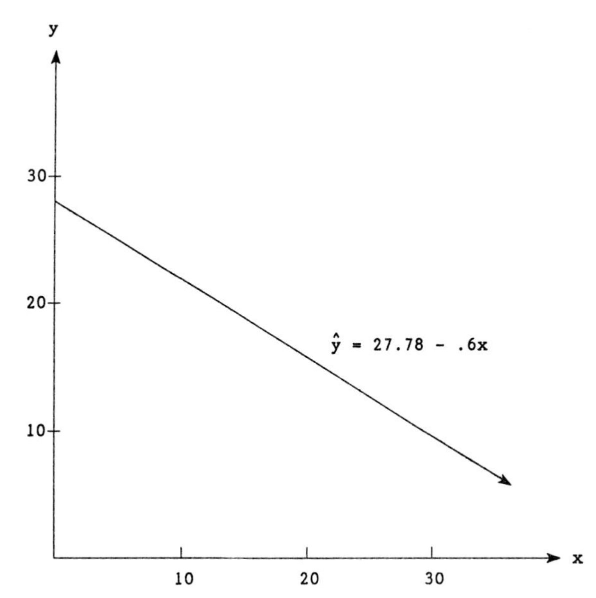

c. $SSE = SS_{yy} - \hat{\beta}_1 SS_{xy} = 25-(-.6)(-30) = 7$

d. $s^2 = \frac{SSE}{n - 2} = \frac{7}{15 - 2} = .5385$

e. For confidence coefficient .90, $\alpha = 1 - .90 = .10$ and $\alpha/2 = .10/2 = .05$. From Table IV, Appendix A, with df = n - 2 = 15 - 2 = 13, $t_{.05} = 1.771$. The 90% confidence interval for β_1 is

$$\hat{\beta}_1 \pm t_{\alpha/2} \frac{s}{\sqrt{SS_{xx}}} \Rightarrow -.6 \pm 1.771 \frac{\sqrt{.5385}}{\sqrt{50}} \Rightarrow -.6 \pm .1838$$

$$\Rightarrow (-.7838, -.4162)$$

We are 90% confident the change in the mean value of y for each unit change in x is between -.7838 and -.4162.

f. For x = 1.8, $\hat{y} = 27.78 - .6(1.8) = 26.7$

The 90% confidence interval is:

$$\hat{y} \pm t_{\alpha/2} s \sqrt{\frac{1}{n} + \frac{(x_p - \bar{x})^2}{SS_{xx}}}$$

$$\Rightarrow 26.7 \pm 1.771(\sqrt{.5385})\sqrt{\frac{1}{15} + \frac{(1.8 - 1.3)^2}{50}}$$

$$\Rightarrow 26.7 \pm .3479 \Rightarrow (26.3521, 27.0479)$$

g. The 90% prediction interval is:

$$\hat{y} \pm t_{\alpha/2} s \sqrt{1 + \frac{1}{n} + \frac{(x_p - \bar{x})^2}{SS_{xx}}}$$

$$\Rightarrow 26.7 \pm 1.771(\sqrt{.5385})\sqrt{1 + \frac{1}{15} + \frac{(1.8 - 1.3)^2}{50}}$$

$$\Rightarrow 26.7 \pm 1.3454 \Rightarrow (25.3546, 28.0454)$$

10.69 a.

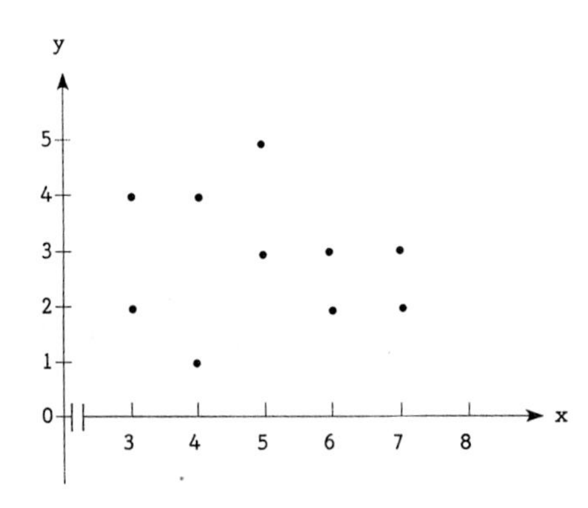

b. Some preliminary calculations are:

$\sum x = 50$ $\sum x^2 = 270$ $\sum xy = 143$

$\sum y = 29$ $\sum y^2 = 97$

$$SS_{xy} = \sum xy - \frac{\sum x \sum y}{n} = 143 - \frac{50(29)}{10} = -2$$

$$SS_{xx} = \sum x^2 - \frac{(\sum x)^2}{n} = 270 - \frac{50^2}{10} = 20$$

$$SS_{yy} = \sum y^2 - \frac{(\sum y)^2}{n} = 97 - \frac{29^2}{10} = 12.9$$

$$r = \frac{SS_{xy}}{\sqrt{SS_{xx} SS_{yy}}} = \frac{-2}{\sqrt{20(12.9)}} = -.1245$$

$$r^2 = (-.1245)^2 = .0155$$

c. Some preliminary calculations are:

$$\hat{\beta}_1 = \frac{SS_{xy}}{SS_{xx}} = \frac{-2}{20} = -.1$$

$$SSE = SS_{yy} - \hat{\beta}_1 SS_{xy} = 12.9-(-.1)(-2) = 12.7$$

$$s^2 = \frac{SSE}{n-2} = \frac{12.7}{10-2} = 1.5875 \qquad s = \sqrt{1.5875} = 1.25996$$

To determine if x and y are linearly correlated, we test:

$H_0\colon \beta_1 = 0$
$H_a\colon \beta_1 \neq 0$

The test statistic is $t = \dfrac{\hat{\beta}_1 - 0}{\dfrac{s}{\sqrt{SS_{xx}}}} = \dfrac{-1.0 - 0}{\dfrac{1.25996}{\sqrt{20}}} = -.35$

The rejection requires $\alpha/2 = .10/2 = .05$ in the each tail of the t distribution with df = n - 2 = 10 - 2 = 8. From Table IV, Appendix A, $t_{.05} = 1.86$. The rejection region is $t > 1.86$ or $t < -1.86$.

Since the observed value of the test statistic does not fall in the rejection region ($t = -.35 \nless -1.86$), H_0 is not rejected. There is insufficient evidence to indicate that x and y are linearly correlated at $\alpha = .10$.

10.71 a. $\hat{\beta}_0 = -13.490347$ - has no meaning since x = 0 is not in the observed range. It is the y-intercept.

$\hat{\beta}_1 = -.052829$ - is the estimated change in the mean proportion of impurity passing through helium for each additional degree.

b. For confidence coefficient .95, $\alpha = .05$ and $\alpha/2 = .025$. From Table IV, Appendix A, with df = n - 2 = 10 - 2 = 8, $t_{.025} = 2.306$. The confidence interval is:

$$\hat{\beta}_1 \pm t_{\alpha/2} s_{\hat{\beta}_1} \Rightarrow -.0528 \pm 2.306(.00772828)$$
$$\Rightarrow -.0528 \pm .0178 \Rightarrow (-.0706, -.0350)$$

We are 95% confident that the change in mean proportion of impurity passing through helium for each additional degree is between -.0706 and -.0350. Since 0 is not in the interval, there is evidence to indicate that temperature contributes information about the proportion of impurity passing through helium.

c. r^2 = R-square = .8538. 85.38% of the total sample variation around the mean proportion of impurity is explained by the linear relationship between proportion of impurity and temperature.

d. From the printout, the 95% prediction interval is (.5987, 1.2653)

10.73 a. $\hat{\beta}_0$(INTERCEP) = -99045

$\hat{\beta}_1$(AREA) = 102.814048

b. To determine if energy consumption is positively linearly related to the shell area, we test:

H_0: $\beta_1 = 0$
H_a: $\beta_1 > 0$

The test statistic is $t = 6.483$ (from printout)

The rejection region requires $\alpha = .10$ in the upper tail of the t distribution with $df = n - 2 = 22 - 2 = 20$. From Table IV, Appendix A, $t_{.10} = 1.325$. The rejection region is $t > 1.325$.

Since the observed value of the test statistic falls in the rejection region ($t = 6.483 > 1.325$), H_0 is rejected. There is sufficient evidence to indicate energy consumption is positively linearly related to the shell area at $\alpha = .10$.

c. Since this is a one tailed test, but the output calculates the p-value for a two tailed test, the observed significance level is

$$\frac{1}{2}(\text{Prob} > |T|) = \frac{1}{2}(.0001) = .00005.$$

d. r^2 = R-SQUARE = .6766

67.66% of the sample variability in energy consumption is explained by the linear relationship between energy consumption and shell area.

e. From the printout, for $x = 8000$, $\hat{y} = 723{,}467$ (observation 23).

The 95% prediction interval is (-631,806, 2,078,740).

This interval is so large and includes negative BTU's; it is not very useful.

10.75 a. $\sum x_i = 206.8$ $\qquad \sum y_i = 206.7$ $\qquad n = 10$

$\sum x_i^2 = 4369.58$ $\qquad \sum x_i y_i = 4355.07$ $\qquad \sum y_i^2 = 4344.33$

$$\bar{x} = \frac{\sum x}{n} = \frac{206.8}{10} = 206.8 \qquad \bar{y} = \frac{\sum y}{n} = \frac{206.7}{10} = 20.67$$

$$SS_{xy} = \sum xy - \frac{(\sum x)(\sum y)}{n} = 4355.07 - \frac{(206.8)(206.7)}{10} = 80.51$$

$$SS_{xx} = \sum x^2 - \frac{(\sum x)^2}{n} = 4369.58 - \frac{(206.8)^2}{10} = 92.96$$

$$\hat{\beta}_1 = \frac{SS_{xy}}{SS_{xx}} = \frac{80.51}{92.96} = .866$$

$\hat{\beta}_0 = \hat{y} - \hat{\beta}_1\bar{x} = 20.67 - .866(20.68) = 2.76$

The least squares equation is:

$\hat{y} = 2.76 - .866x$

b. $SS_{yy} = \sum y^2 - \frac{(\sum y)^2}{n} = 4344.33 - \frac{(206.7)^2}{10} = 71.84$

$$r = \frac{SS_{xy}}{\sqrt{SS_{xy}SS_{yy}}} = \frac{80.51}{\sqrt{(92.96)(71.84)}} = .985$$

$r^2 = (.985)^2 = .971$

r = .985 implies there is a fairly strong positive linear relationship between the reliable method and the new instrument.

$r^2 = .971$. 97.1% of the sample variability of the new instrument is explained by the linear relationship between the new instrument and the reliable method.

c. The form of the prediction interval is

$$\hat{y} \pm t_{\alpha/2}s\sqrt{1 + \frac{1}{n} + \frac{(x_p - \bar{x})^2}{SS_{xx}}}$$

where $\hat{y} = 2.76 + .866(20.0) = 20.08$.

For confidence coefficient .90, $\alpha = .10$ and $\alpha/2 = .05$. From Table IV, Appendix A, with df = n - 2 = 10 - 2 = 8, $t_{.05} = 1.860$.

We also need to calculate

$$s^2 = \frac{SS_{yy} - \hat{\beta}_1 SS_{xy}}{n - 2} = \frac{71.84 - (.866)(80.51)}{10 - 2} = .2648$$

Then, the standard error of this regression is

$s = \sqrt{s^2} = \sqrt{.2648} = .5146$.

The prediction interval is

$$20.08 \pm 1.860(.5146)\sqrt{1 + \frac{1}{10} + \frac{(20.0 - 20.68)^2}{92.96}}$$

$\Rightarrow 20.08 \pm 1.01 \Rightarrow (19.07, 21.09)$

10.77 a. Some preliminary calculations are:

$\sum x = 9.5 \qquad \sum x^2 = 34.75 \qquad \sum xy = 496.925$

$\sum y = 118.92 \qquad \sum y^2 = 7318.2812$

$$SS_{xy} = \sum xy - \frac{\sum x \sum y}{n} = 496.925 - \frac{9.5(118.92)}{4} = 214.49$$

$$SS_{xx} = \sum x^2 - \frac{(\sum x)^2}{n} = 34.75 - \frac{9.5^2}{4} = 12.1875$$

$$SS_{yy} = \sum y^2 - \frac{(\sum y)^2}{n} = 7318.2812 - \frac{118.92^2}{4} = 3782.7896$$

$$\hat{\beta}_1 = \frac{SS_{xy}}{SS_{xx}} = \frac{214.49}{12.1875} = 17.59917949 \approx 17.599$$

$$\hat{\beta}_0 = \bar{y} - \hat{\beta}_1 \bar{x} = \frac{118.92}{4} - 17.59917949\left(\frac{9.5}{4}\right) = -12.0681$$

$$SSE = SS_{yy} - \hat{\beta}_1 SS_{xy} = 3782.7896 - 17.59917949(214.49) = 7.941591$$

$$s^2 = \frac{SSE}{n-2} = \frac{7.941591}{4-2} = 3.9708$$

b. For confidence coefficient .95, $\alpha = .05$ and $\alpha/2 = .025$. From Table IV, Appendix A, with df = n - 2 = 4 - 2 = 2, $t_{.025} = 4.303$.

The confidence interval is:

$$\hat{\beta}_1 \pm t_{\alpha/2} \frac{s}{\sqrt{SS_{xx}}} \Rightarrow 17.599 \pm 4.303 \frac{\sqrt{3.9708}}{\sqrt{12.1875}}$$

$$\Rightarrow 17.599 \pm 2.4561 \Rightarrow (15.1429, 20.0551)$$

c. We must assume the random error terms (ε_i's) are normally distributed with a mean of 0 and common variance σ^2. Also, the random error terms must be independent.

d. For x = 4, $\hat{y} = -12.0681 + 17.599(4) = 58.3279$

The 95% confidence interval is:

$$\hat{y} \pm t_{\alpha/2} s \sqrt{\frac{1}{n} + \frac{(x_p - \bar{x})^2}{SS_{xx}}}$$

$$\Rightarrow 58.3279 \pm 4.303(\sqrt{3.9708})\sqrt{\frac{1}{4} + \frac{(4 - 2.375)^2}{12.1875}}$$

$$\Rightarrow 58.3279 \pm 5.8575 \Rightarrow (52.4704, 64.1854)$$